Security Informatics

Annals of Information Systems

Volume 1:
Managing in the Information Economy: *Current Research Issues*
Uday Apte, Uday Karmarkar, *eds.*

Volume 2:
Decision Support for Global Enterprises
Uday Kulkarni, Daniel J. Power and Ramesh Sharda, *eds.*

Volume 3:
New Trends in Data Warehousing and Data Analysis
Stanislaw Kozielski, Robert Wremble, *eds.*

Volume 4:
Knowledge Management and Organizational Learning
William R. King, *ed.*

Volume 5:
Information Technology and Product Development
Satish Nambisan, *ed.*

Volume 6:
Web 2.0 & Semantic Web
Vladan Devedžic, Dragan Gašević, *eds.*

Volume 7:
Web-Based Applications in Healthcare and Biomedicine
Athina Lazakidou, *ed.*

Volume 8:
Data Mining: Special Issue in Annals of Information Systems
Robert Stahlbock, Sven F. Crone and Stefan Lessmann, *eds.*

Volume 9:
Security Informatics
Christopher C. Yang, Michael Chiu-Lung Chau, Jau-Hwang Wang
and Hsinchun Chen, *eds.*

Christopher C. Yang · Michael Chiu-Lung Chau ·
Jau-Hwang Wang · Hsinchun Chen
Editors

Security Informatics

 Springer

Editors
Christopher C. Yang
College of Information Science and
    Technology
Drexel University
3141 Chestnut Street
Philadelphia, PA 19104
USA
chris.yang@drexel.edu

Jau-Hwang Wang
Department of Information
    Management
Central Police University
Gueishan
Taoyuan 33334
Taiwan R.O.C
jwang@mail.cpu.edu.tw

Michael Chiu-Lung Chau
School of Business
7/F, Meng Wah Complex
The University of Hong Kong
Pokfulam Road
Hong Kong SAR
mchau@business.hku.hk

Hsinchun Chen
Eller College of Management
McClelland Hall 430
1130 E. Helen St.
Tucson, AZ 85721
USA
hchen@eller.arizona.edu

ISSN 1934-3221   e-ISSN 1934-3213
ISBN 978-1-4419-1324-1   e-ISBN 978-1-4419-1325-8
DOI 10.1007/978-1-4419-1325-8
Springer New York Dordrecht Heidelberg London

Library of Congress Control Number: 2009938002

Printed on acid-free paper

Springer is part of Springer Science+Business Media (www.springer.com)

Preface

Intelligence and Security Informatics (ISI) is the study of the development and use of advanced information systems and technologies for national, international, and societal security-related applications. ISI topics include ISI data management, data and text mining for ISI applications, terrorism informatics, deception and intent detection, terrorist and criminal social network analysis, public health and biosecurity, crime analysis, cyber-infrastructure protection, transportation infrastructure security, policy studies and evaluation, and information assurance. Due to the advent of terrorism attacks in recent years, ISI has gained increasingly more interest and attention from academic researchers, law enforcement and intelligence experts, information technology consultants, and practitioners. The first IEEE International Conference on ISI was held in 2003 and subsequent meetings were held annually. Starting from 2006, the Pacific Asia Workshop on ISI (PAISI) was also held annually in Pacific Asia, with large numbers of contributors and participants from the region. A European Workshop on ISI was started in Denmark in 2008. Given the importance of ISI, we introduce a special volume of security informatics in *Annals of Information Systems*.

In this special volume, we include nine papers covering a wide range of active research areas in security informatics. Chen et al. studied the international Falun Gong movement through a cyber-archaeology approach. Link, web content, and forum content analyses were employed to investigate the framing of social movement identity. Skillicorn and Little adopted an empirical model based on changes in frequencies of 88 significant words to detect deception in non-emotional testimony. The model was validated with the testimony to the Gomery Commission, a Canadian investigation of misuse of government funds. Yang and Tang developed a new sub-graph generalization approach for social network data sharing between organizations and preserving privacy of sensitive data. The shared data was integrated to support social network analysis such as centrality measurement.

Qiao et al. proposed the Growing Window-based Constrained k-Closest Pairs (GWCCP) algorithm to discover the closest pairs of objects within a constrained spatial region for spatial analysis in crime databases. Lee et al. developed a framework for what-if emergency response management using a complete set of order-k, ordered order-k and kth nearest Voronoi diagram. Several scenarios had been presented to demonstrate how it was useful in the four phases of emergency response: mitigation, preparedness, response, and recovery.

Glässer and Vajihollahi proposed a framework of identity management architecture for analyzing and reasoning about identity management concepts and requirements. Chen et al. proposed a framework of two-staged game theoretic models for deploying intrusion detection agents. Su et al. presented the distributed event-triggered knowledge network (ETKnet) for data and knowledge sharing. A unified knowledge and process specification language was also developed. Cha et al. proposed a risk management system RiskPatrol which integrated the business continuity management with the risk assessment process.

Philadelphia, Pennsylvania	*Christopher C. Yang*
Hong Kong, SAR	*Michael Chau*
Taiwan, ROC	*Jau-Hwang Wang*
Tucson, Arizona	*Hsinchun Chen*

Contents

Framing Social Movement Identity with Cyber-Artifacts: A Case Study of the International Falun Gong Movement

Yi-Da Chen, Ahmed Abbasi, and Hsinchun Chen

Abstract Framing a collective identity is an essential process in a social movement. The identity defines the orientation of public actions to take and establishes an informal interaction network for circulating important information and material resources. While domestic social movements emphasize the coherence of identity in alliance, global or cyber-activism is now flexible in its collective identity given the rise of the Internet. A campaign may include diverse social movement organizations (SMOs) with different social agendas. This flexible identity framing encourages personal involvement in direct action. On the other hand, it may damage solidarity within SMOs and make campaigns difficult to control. To assess the sustainability of an SMO, it is important to understand its collective identity and the social codes embedded within its associated cyber-societies and cyber-artifacts. In this study, we took a cyber-archeology approach and used the international Falun Gong (FLG) movement as a case study to investigate this identity-framing issue. We employed social network analysis and Writeprint to analyze FLG's cyber-artifacts from the perspectives of links, web content, and forum content. In the link analysis, FLG's websites linked closely to Chinese democracy and human rights SMOs, reflecting FLG's historical conflicts with the Chinese government after the official ban in 1999. In the web content analysis, we used Writeprint to analyze the writings of Hongzhi Li and of his editors, and found that Hongzhi Li's writings center around the ideological teaching of Falun Dafa while the editors post specific programs to realize Li's teaching. In the forum content analysis, FLG comprehensively organizes several different concepts on a continuum: from FLG ideology to life philosophy and mysterious phenomena, and from mysterious phenomena to anti-Chinese Communist

Yi-Da Chen, Hsinchun Chen*
Artificial Intelligence Lab, Department of Management Information Systems, University of Arizona, Tucson, AZ 85721, USA
e-mail: ydchenb@email.arizona.edu; *hchen@eller.arizona.edu

Ahmed Abbasi
Sheldon B. Lubar School Business, University of Wisconsin-Milwaukee, Milwaukee, WI 53201, USA
e-mail: abbasi@uwm.edu

C.C. Yang et al. (eds.), *Security Informatics*, Annals of Information Systems 9, DOI 10.1007/978-1-4419-1325-8_1, © Springer Science+Business Media, LLC 2010

Party and persecution by conceptualizing the Chinese government as the Evil. By deploying those cyber-artifacts, FLG seamlessly connects different ideologies and establishes its identity as a Qi-Gong, religious, and activist group.

Keywords Social movement · Collective identity · Falun Gong · Internet · Social network analysis · Writeprints

1 Introduction

The Internet nowadays is not merely a digital platform for exchanging data or information but is also a communication channel for individuals to share and promulgate their beliefs and ideas. People discuss topics of interest in forums or create their own blogs for posting experiences and opinions. Through hyperlinks, individuals can link their own posts to other web pages for citing the comments with which they agree or oppose. Gradually, "ideological networks" are formed on the Internet in which websites with similar ideas are connected together via hyperlinks. One may find a number of relevant websites of her/his interests by just following the hyperlinks of a few seed websites.

The Internet has also changed how social movement organizations (SMOs) advocate their ideology and mobilize their resources. They are no longer restricted by time or space. Their websites hold permanent campaigns appealing to a global audience [6, 7, 17]. The low cost of communication makes them more easily align together for large public actions [6, 7, 25]. For example, a demonstration against the war in Iraq in Washington, D.C., 2003, gathered around 100,000 people with different protest positions [7]. Furthermore, activist groups can increase individual participation by hyperlinking an inclusive ideological network which provides multiple entry points for potential supporters to join in [6]. The anti-Microsoft network, for instance, involves a great diversity of interest groups, including corporations, consumer protection organizations, and labor alliances [6].

In the traditional social movement theory, framing a coherent identity is a crucial process for an SMO to establish itself in social movements [19]. The collective identity helps participants to develop a trust relationship with their members and creates an informal interaction network for circulating important information and material resources [29]. The failure to achieve common framing can create tension and fragmentation within coordinated SMOs [7]. Global or cyberactivism, conversely, is flexible in identity framing [7]. Its advocate network may contain diverse social justice agendas, rich in personal appeal but thin in ideology [6, 7]. The ease of individuals joining and leaving a given network damages solidarity in SMOs and makes campaigns difficult to control [6]. To assess the sustainability of an SMO in the Internet era, it is important to understand how the SMO constructs identity and social codes within its associated cyber-societies [7].

In this study, we took a cyber-archaeology approach to investigate the framing of collective identity on the Internet and used the International Falun Gong (FLG)

movement as a case study. The cyber-archaeology approach adapts methods from archaeology and anthropology to analyze a community's cultural cyber-artifacts [16, 27]. In applying the cyber-archaeology approach to the study of SMOs, the approach involves (1) the identification of their associated cyber-societies, (2) collection of cyber-artifacts with automated procedures, and (3) analysis of cyber-artifacts from the perspective of the social movement theory. The FLG was chosen as our case study because it involves various identities, including Qi-Gong exercises, new religion, and activism [28, 30, 34, 21], and heavily uses the Internet as a vehicle for information dissemination [5]. Our goal was to investigate how FLG comprehensively maintains these three identities simultaneously within its cyber-societies.

The remainder of the study is organized as follows. We first review social movement theory and cyber-archaeology, and introduce two analytical tools used for examining cyber-artifacts: social network analysis (SNA) and Writeprint. Then we describe our research design which was based on the cyber-archaeology research framework, covering link, web content, and forum content analyses. Finally, we present our results and conclusions.

2 Literature Review

2.1 Social Movement Theory

Social movements are often recognized as irrational and disorganized activities working against social injustices. Unlike official political participation or lobbying, activists organize a broad range of individuals and seek to build a radically new social order through public protest [11]. However, the study of social movements reveals that they have much deeper organizational and psychological foundations than would appear. Activists do not just blindly deploy protest actions: they calculate the costs and benefits of the actions with present resources before initiation [29]. People are not irrationally supporting a social movement: they seek an identity and share a sense of belonging through their participation [29].

Collective action and resource mobilization are two intellectual currents which dominated the early development of social movement theory [19, 29]. Collective action is based on the idea that social movements are triggered by an overly rapid social transformation. In this school of thought, a society consisted of several balanced sub-systems and a movement reflective of the failure of social institutions or control mechanisms to rebalance them after a dramatic change [29]. In such a moment, the existing norms no longer provide a sufficient rationale for social action, and a group of individuals sees the situation as injustice and reacts to it by developing shared beliefs and new norms for behavior. The American Civil War and Civil Rights Movement are significant examples illustrating this point of view [29]. The framing of collective identity is an essential process for collective action [19, 29, 20]. For participants, the identity helps establish a trust relationship with others in the same group and excludes those whom they oppose [19, 29]. For the movement, it defines the orientation of public action and constrains where the actions will take place [19, 20].

Resource mobilization examines the strategic components in social movements. It is based on two main assertions: (1) movement activities are not spontaneous and disorganized and (2) their participants are not irrational [24]. In other words, social movements are meaningful actions with specific purposes [29]. In this school of thought, the movements involve so-called "social movement entrepreneurs" who call public attention to problems and who recruit and mobilize supporters for action [19]. Public protests are derived from a careful calculation of costs and benefits as measured against present resources [29]. To achieve their goals, activists need to control the costs of their actions, organize discontent to reach social consensus, and create solidarity networks for circulating information and resources. From this point of view, social movements are an extension of formal political actions to pursue social interests [29].

2.2 Social Movement Organizations and the Internet

SMOs are non-governmental organizations which seek to realize the ideology of a social movement with clear goals [29]. Their coordination in a movement is suggested following a SPIN model: (1) Segmented, (2) Polycephalous, (3) Integrated, and (4) Networked [6, 29, 14]. According to the model, a social movement is composed of many SMOs which continuously die or rise [29]. It has multiple and sometimes competing leaders or influential centers [6, 14]. Those SMOs form a loose but integrated network with overlapping membership and shared ideas [6, 29, 14].

The Internet has changed the ways SMOs operate [6, 7, 25]. One advantage that the Internet brought to SMOs is the reduction of communication costs [6, 25]. In the pre-Internet era, SMOs relied on informal interaction networks of their supporters to circulate information and mobilize participants for actions [29]. They now use Computer-Mediated Communication (CMC), including e-mails and forums, to promote their ideology, recruit new members, and coordinate campaigns. Another significant change is that activists can hold permanent campaigns via their websites, thereby appealing to a global audience and transforming domestic social movements into global or cyber-activism [6, 7, 23]. Compared to traditional activism, global activism depends on the availability of technology networks in order to expand, involves diverse social justice agendas and multiple issues, and promotes personal involvement in direct action [7]. To reflect this transformation, Gerlach revised his SPIN model from "polycephalous" to "polycentric," meaning that global activist networks have many centers or hubs for supporters to join or leave, and are less likely to be managed by permanent leaders [6]. Therefore, they are thin in ideology but rich in personal identity and lifestyle narratives [6]. To assess the sustainability and quality of an SMO in the global movement, it is crucial to identify the social codes and values embedded in its CMC cyber-artifacts [7].

2.3 Cyber-Society Archaeology

Researchers have recognized the potential for conducting research grounded in social movement theory through analysis of SMO cyber-societies [10, 33]. In a cyber-society, all cultural artifacts of the community are recorded and preserved, and can be recovered by the social scientist if desired. The discourse among community members is rich, robust, filled with social cues, and highly revealing of individual perspectives. Traditional studies in archaeology and anthropology face the difficult task of determining the time associated with collected artifacts, but in an online environment, time is often measured and recorded precisely for each cyber-artifact. The accessibility of time information allows for analysis of the exact evolution of interactions and the production of cultural artifacts, which can reveal a great deal about social behavior.

Social movement research on cyber-societies requires adaptation of the traditional methods of studying conventional societies to virtual communities. Cyber-society researchers advocate using methods adapted from archaeology and anthropology to study a community's cultural cyber-artifacts [16, 27]. The application of methods used in archaeology to cyber-societies has been termed cyber-archaeology, and focuses on the analysis of CMC cyber-artifacts [15]. In particular, researchers have proposed the use of webometric methodologies in cyber-archaeology studies for the collection of cyber-artifacts [8]. Forms of analysis associated with webometrics include web content, link, technology, and usage analysis [4, 9]. Web content analysis focuses on the CMC communications and textual information provided on a web page. Link analysis considers the network of websites linked to and from a particular website. The sophistication of the platform for CMC communication is studied through web technology analysis. Web usage analysis focuses on user behavior and activity logs.

A systematic cyber-archaeology framework is proposed here to guide social movement research on SMO cyber-societies. The framework shown in Fig. 1 leverages traditional archaeology and anthropology research approaches and perspectives, as suggested by cyber-society researchers [16, 27]. CMC cyber-artifacts are collected, categorized, and visualized using webometric approaches [8]. The proposed framework for cyber-archaeology has three phases. In phase one, social movement research design, social researchers identify SMOs and cyber-societies of interest, and target cyber-artifacts for collection and analysis. Phase two consists of the automated collection and classification of cyber-artifacts across one or many identified cyber-societies. The third phase supports the analysis of cyber-artifacts by social researchers from the perspectives of social movement theory [22].

2.4 Social Network Analysis

The central theme in social movement studies revolves around how activists "organize" themselves in campaigns to achieve impact on governments and societies,

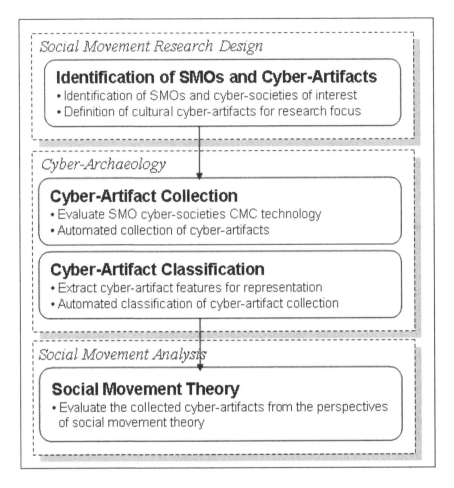

Fig. 1 Cyber-archaeology framework

which makes SNA perfectly suitable for this kind of investigation. SNA is a graph-based methodology used to analyze social relations and their influence on individual behavior and organizational structure. It was developed by sociologists and has been applied in several academic fields, such as epidemiology and CMC [18]. In SNA, individuals or actors are represented as nodes in a simple network graph, called a social network, and tied with edges indicating relationships. The visualization of a social network can provide a basic sense of how actors affiliate with each other and what their roles are in the group. In a sample friendship network, as shown in Fig. 2, Persons A and G are considered the most active or "popular" persons since they are linked to the largest number of people. Person F is also important although he/she doesn't have as many connections as Persons A and G: Person F bridges two

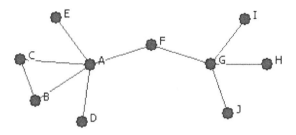

Fig. 2 Sample friendship network

different clusters of friends. In SNA, these three people are said to be central or prominent within the sample network.

Centrality measures are quantitative indicators for finding those "central" individuals from a network originally developed in communication scenarios. From a topological perspective, people who are able to receive or control the mainstream of message flow typically stand in a position similar to the central point of a star [12], such as the location of Person A in the sample network. Various centrality measures, such as degree and betweenness, can be employed to determine the importance of a node within a network. Degree is the number of edges that a node has. Since the central point of a star has the largest number of edges connecting it to the other nodes, a node with a higher degree is topologically considered to be more central to its network [12, 35]. Betweenness measures "the extent to which a particular node lies between the various other nodes" [32] because the central point also sits between pairs. The higher betweenness a node has, the more potential it has to be a gatekeeper controlling the connections (such as communications) between the others [32].

Through reconstructing the network of nineteenth-century women reform leaders in New York State, Rosenthal et al. [31] reported that weak ties played an important role in the women's movement: while strong ties linked major women's organizations in a movement, weak ties bridged several clusters and channeled the communication to diverse audiences. After the emergence of global activism, many researchers shifted their focus from SMO physical connections to SMO website hyperlinks. Ackland et al. [3] used their VOSON system to demonstrate the usefulness of network visualization in the analysis of linkage between environmental groups on the Internet. Garrido and Halavais [13] used hyperlink analysis to map the Zapatista online network and examine its affiliation with other SMOs. They found that the secondary tier of Zapatista related websites played a bridging role linking the Zapatista network to the global SMO network.

2.5 Writeprints

Because of its anonymous nature, the Internet has become a major medium for cybercrime ranging from illegal sales and phishing to terrorist communication. In

order to increase the awareness and accountability of users, many studies have been
devoted to developing techniques to identify authors in the online environment [1].
Authorship identification is a process of matching unidentified writings to an author
based on the similarity of writing styles between the known works of the author
and unidentified pieces [1, 2]. Four major categories of style features have been
extensively used to identify writing styles: lexical, syntactic, structural, and content-
specific [1]. Lexical features include total number of words, words per sentence, and
word length distribution. Syntax refers to the patterns used for the formation of sen-
tences, such as punctuation and function/stop words. Structural features deal with
the organization and layout of the text, such as the use of greetings and signatures,
the number of paragraphs, and average paragraph length. Content-specific features
are keywords that are important within a specific topic domain. Among these four
categories, lexical and syntactic features are frequently used because of their high
discriminatory ability and portability across domains [1, 2].

In 2006, Abbasi and Chen [2] proposed a visualization technique for authorship
called Writeprints as shown in Fig. 3, which is analogous to the fingerprint biomet-
ric system. Unlike other studies of authorship visualization merely using n-gram
features for discrimination, Writeprint was designed to apply to a large number of
authors in an online setting. It uses all four major types of style features: lexical, syntactic, structural, and content-specific [2]. The generation of a "Writeprint" consists
of two main steps: (1) reduce dimensionality and (2) create visualizations. After
extracting features from a set of documents, the Writeprint adopts principal component analysis (PCA) to reduce dimensionality of feature usage vectors by using
the two principal components or eigenvectors with the largest eigenvalues. Once the
eigenvectors have been computed with PCA, a sliding window algorithm is used
to extract the feature usage vector for the text region inside a window, which slides
over the text. For each window instance, the sum of the product of the principal component (primary eigenvector) and the feature vector represents the x-coordinate of
the pattern point while the sum of the product of the second component (secondary
eigenvector) and the feature vector represents the y-coordinate of the data point.
Each data point generated is then plotted onto a two-dimensional space to create the

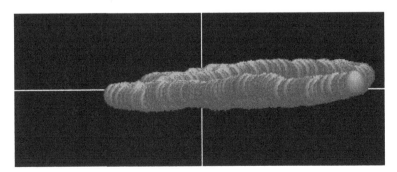

Fig. 3 Example of the Writeprint (each dot represents a text segment overlaying on the X–Y-
coordinates)

Writeprint. Abbasi and Chen [2] reported that Writeprints outperformed the Support Vector Machine (SVM) in classification of online messages in their evaluation.

3 Research Design: A Case Study of the International FLG Movement

This study employed the proposed cyber-archaeology framework and used the Falun Gong (FLG, 法輪功) movement as a case study to investigate SMOs' collective identity on the Internet and their ideological ties with others.

3.1 The Falun Gong Movement

FLG was founded by Hongzhi Li (李洪志) and introduced to the public in Mainland China in 1992 [28]. It focuses on the concept of cultivation and has two main components: practice and principle. The practice component includes five sets of gentle exercise movements which are similar to Qi-Gong (氣功) exercises. The principal component emphasizes the importance of truthfulness, compassion, and forbearance (真善忍). FLG practitioners believe they can enhance their physical health and elevate their mind at the same time.

On July 20, 1999, the Chinese government designated FLG as an evil cult and banned all its public activities [34]. This suppression is widely believed to be related to the mass petition of Falun Gong practitioners in Zhongnanhai, Beijing on April 25, 1999 [28]. After the official ban, Hongzhi Li stayed in the United States and used FLG websites, such as Clearwisdom.net (法輪大法明慧網), to release his articles and continue his teaching. Currently, FLG has local websites of practitioners in over 38 countries and five continents. It holds several conferences annually in North America and Europe.

Before the suppression, there was no evidence that FLG or Hongzhi Li had any political agenda against the Chinese government. On January 23, 2001, a self-immolation incident by five FLG practitioners in Tiananmen Square in Beijing was reported by the international news media. In 2002, FLG websites began releasing accounts of persecution against practitioners in Mainland China, including pictures, stories, and persecution methods. In late 2004, the Epoch Times (大紀元), which is related to FLG, published "Nine Commentaries on the Communist Party" (九評共產黨) and held a "Quitting the CCP (Chinese Communist Party)" (退黨) campaign.

Several studies, each using a different approach, have investigated how FLG transformed from a simple Qi-Gong exercise group to a religious and social movement organization [28, 30, 34, 21]. Lu [21] applied the religious economy model to the formation of new religions to interpret the shift of FLG from a healing system to a religion. According to Lu's analysis [21], Hongzhi Li purposely introduced his own theory of salvation, Falun Dafa, to differentiate FLG from other competing Qi-Gong groups. In addition, he used various organizational and doctrinal mechanisms to keep his practitioners and avoid schisms. For example, he claimed that he was the incarnation of the highest supernatural force and that the only master in Dafa [21]. Rahn [30] used the conflict between FLG and the Chinese government to explain FLG's role in social movements. He suggested that the persecution of FLG in China was the key to establishing FLG's identity as a human rights movement. By examining Hongzhi Li's messages in the Clearwisdom.net, he brought out a concern that the Fa-rectification teaching may induce violent behavior in FLG practitioners, because the frustration at achieving this ultimate goal may intensify "the battle between good and evil" [30]. But Rahn [30] asserted that FLG's positive public image as a human rights group can decrease the chances of a violent act.

The widespread use of the Internet for organizing practitioners is another research focus of those studying FLG. Bell and Boas [5] summarized three important functions of the Internet in the FLG movement: (1) disseminating Hongzhi Li's teachings, (2) strengthening the integrity of a globally dispersed community, and (3) bringing pressure on the Chinese government for lifting the ban. They also concluded that the use of the Internet might bring splinter sects challenging Li's authority [5]. This is consistent with Bennett's point of view that "the ease of joining and leaving polycentric issue networks means that it becomes difficult to control campaigns or to achieve coherent collective identity frames" [6].

3.2 Research Design

Figure 4 shows our research design. Our interest was in two types of cyber-artifacts in the FLG movement: websites and forums. SMO websites typically are official "entry points" of SMO networks advocating their ideologies and campaigns. Collecting those websites' hyperlinks and contents allowed us to investigate how FLG officially deploys itself on the Internet and connects to other SMOs. SMO forums, on the other hand, provide a relatively intimate view of how members interact with each other and discuss their SMO ideology. We used automatic programs to collect those cyber-artifacts and performed three analyses for our inquiries: link, web content, and forum content analyses.

3.2.1 Cyber-Artifacts Collection

Website Hyperlinks and Content: FLG has four core websites as listed in Table 1, distributing FLG news, Hongzhi Li's articles, and accounts of persecution of FLG practitioners in Mainland China. These four core websites are Clearwisdom.net (法輪大法明慧網), FalunInfo.net (法輪大法新聞社), FalunDafa. org (法輪大法), and EpochTimes.com (大紀元). Each website offers more than ten language versions and has multiple domain names, a reflection of the fact that the FLG movement is organized on a global scale.

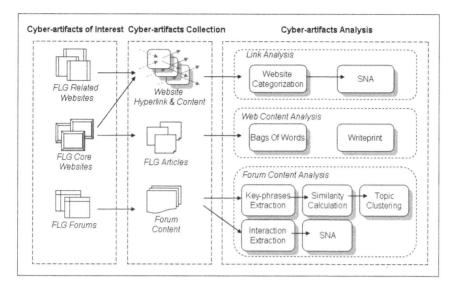

Fig. 4   Research design

Table 1   Four core FLG websites

Website	Content	Language	Domain name
Clearwisdom.net (法輪大法明慧網)	1.Falun Gong Information 2.Hongzi Li' Articles 3.Persecution Accounts 4.Practitioners'Sharing	10	27
FalunDafa.org (法輪大法)	1.Falun Gong Information 2.Local Contact Information and Websites	36	4
FalunInfo.net (法輪大法新聞社)	1.Persecution Accounts 2.Falun Gong News	12	14
EpochTimes.com (大紀元)	1.World News 2.Persecution Accounts 3.CCP Criticism	15	36

We automatically collected (spidered) FLG relevant websites, including those of other activist groups having hyperlinks to FLG websites, with two levels of inlinks and outlinks via 31 seed websites, which included the four core websites and another 27 FLG websites identified by Google search. A total of 425 relevant domain names were found during spidering and 172 were deemed relevant. Most of the relevant websites were found to be directly linked to the core websites as shown in Table 2.

FLG Articles: For the web content, we were particularly interested in studying the role of Hongzhi Li's articles in the FLG movement. From Clearwisdom.net, we collected 135 articles from Hongzhi Li and 74 articles from the editors for later

Table 2 Collecting FLG-related websites via seed websites

Seed Website	Outlink Level 1	Outlink Level 2	Inlink Level 1	Inlink Level 2	Total
FalunDafa.org (法輪大法)	85	3	3	0	91
EpochTimes.com (大紀元)	26	0	1	1	28
Clearwisdom.net (法輪大法明慧網)	15	4	1	1	21
GuangMing.org (澳洲光明網)	12	0	1	0	13
FalunInfo.net (法輪大法新聞社)	4	2	1	0	7
SoundOfHope.org (希望之聲電台)	2	2	0	0	4
GrandTrial.org (全球公審江澤民)	3	0	0	0	3
GlobalRescue.net (全球營救 FLG 學員)	2	0	0	0	2
ZhuiChaGouJi.org (追查迫害 FLG 組織)	0	1	0	0	1
NtdTV.com (新唐人電視台)	0	1	0	0	1
Minghui-School.org (明慧學校)	1	0	0	0	1
Total	150	13	7	2	172

comparison. Those articles concentrated on the discussion of three topics: teaching/principles of FLG, the position of FLG on political issues especially related to Mainland China, and summaries of various FLG conferences. The summary of these two sets of articles is shown in Table 3. Compared to the editors' articles, Hongzhi Li's are much longer in length.

Table 3 Summary of articles in Clearwisdom.net

Source	Number of articles	Words per article	Duration
Hongzhi Li	135	1,430	5/1999–2/2007
Editors	74	670	3/2000–12/2006
Total	209	1,161	5/1999–2/2007

Forum Content: We used Google search and website linkage to find FLG forums. Four forums were found, but only one forum, Falun Dafa Universal (世界法輪大法研究會), was, and is, still active and has more than 50 authors contributing to it. Therefore, we concentrated on the analysis of threads and messages in this forum. Falun Dafa Universal, located at city.udn.com (網路城邦), was established in 2005. It has 120 members and 28 discussion boards covering Hongzhi Li's articles, persecution accounts, and the FLG universal and science database. This forum circulates many articles from the four core websites. Thus, the average length of messages is long, 1,288 characters per message, but the average reply rate is low, 0.89 reply messages per thread. A total of 740 threads and 1,399 messages were collected for this forum.

3.2.2 Cyber-Artifact Analysis

Link Analysis: In order to understand the main ideas of these websites and how they linked together, we first classified their ideological types and performed SNA to analyze their network structure. Two measures of centrality in SNA are used to investigate which websites are prominent in this network: degree and betweenness. The degree of a node is the number of links it has, reflecting its activity level. Betweenness is a measure of the frequency with which a node lies on the shortest geodesic paths of pairs of other nodes. It can be used to detect the extent to which a node plays the role of a gatekeeper in controlling the communication of others [32].

Web Content Analysis: In order to highlight the characteristics of Hongzhi Li's writing, we used Bag of Words and Writeprints, developed by Abbasi and Chen [2], and compared his articles with other articles written by the editors of Clearwisdom.net.

Forum Content Analysis: At the forum level, we performed two types of analysis: thread topic and author interaction. In the thread topic analysis, we investigated how many topics are covered in this forum and how those topics relate to each other. Since Falun Dafa Universal is a Chinese forum, we first used MI, a Chinese phrase extraction tool developed by Ong and Chen [26], to extract key Chinese phrases from the threads and convert those threads into vectors of those key phrases. The top 20 key phrases based on frequency of appearance are shown in Table 4. We then used the cosine coefficient to calculate the similarity between threads and displayed those threads in a two-dimensional map. For author interaction analysis, we extracted the authors' responses to others' threads and performed SNA based on their interaction history.

4 Research Results

In this section, we present our research results.

4.1 Link Analysis

The 203 FLG relevant websites, including seed websites and those identified via the seeds, are classified into five main categories based on their web content: FLG

Table 4 Key Chinese phrases of forum messages

Rank	Phrase	Rank	Phrase
1	法輪 (Falun)	11	政府 (Goverment)
2	法輪功 (Falun Gong)	12	醫院 (Hospital)
3	中國 (China)	13	修煉 (Cultivation)
4	學員 (Practitioner)	14	問題 (Problem)
5	器官 (Organ)	15	國際 (International)
6	迫害 (Persecution)	16	人類 (Human)
7	美國 (the United States)	17	個人 (Individual)
8	蘇家屯 (Sujiatun Camp)	18	國家 (Country)
9	大法 (Dafa)	19	集中營 (Labor Camp)
10	社會 (Society)	20	人民 (People)

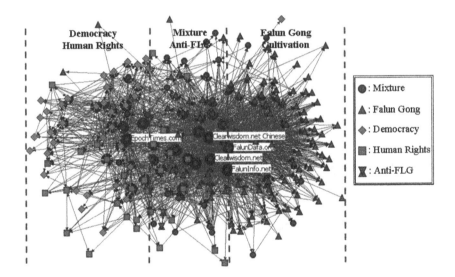

Fig. 5 Network of FLG relevant websites

cultivation, human rights, democracy, anti-FLG, and mixture (topics of more than one category). Two coders were hired for the website classification. The network of these websites, displayed with a spring embedded algorithm, is shown in Fig. 5. The network has three main components: human rights and democracy on the left-hand side, FLG cultivation on the right-hand side, and mixture (and anti-FLG) in the middle. The mixture websites, including Clearwisdom.net and EpochTimes.com, act as bridges connecting the other two main components. The human rights and democracy websites are somewhat mixed together.

We used two centrality measures, degree and betweenness, to identify the most prominent websites within this network. Here, the degree or in-degree is calculated by the number of inlinks and reflects the popularity of a website. The betweenness measures the potential that a website may be a gatekeeper controlling the interaction with other websites. The top ten prominent websites in this network are listed in Table 5. The four core FLG websites are at the top of the list.

Table 5 Top ten websites in FLG network based on centrality

Rank	Website	In Degree	Website	Betweenness
1	FalunDafa.org (法輪大法)	113	FalunDafa.org (法輪大法)	14657.33
2	FalunInfo.net (法輪大法新聞社)	99	EpochTimes.com (大紀元)	6166.15
3	Clearwisdom.net (法輪大法明慧網英文)	90	Clearwisdom.net Chinese version (法輪大法明慧網中文)	4318.54
4	Clearwisdom.net Chinese version (法輪大法明慧網中文)	88	GuangMing.org (澳洲光明網)	2533.59
5	EpochTimes.com (大紀元)	78	Clearwisdom.net (法輪大法明慧網英文)	2298.31
6	ZhengJian.org (正見)	65	FalunInfo.net (法輪大法新聞社)	2014.37
7	Fofg.org (法輪功之友)	54	ZhengJian.org (正見)	1335.54
8	ClearHarmony.net (歐洲圓明網)	50	SoundOfHope.org (希望之聲電台)	1276.49
9	SoundOfHope.org (希望之聲電台)	50	ClearHarmony.net (歐洲圓明網)	1077.36
10	NtdTV.com (新唐人電視台)	48	HriChina.org (中國人權)	792.51

We used the inlinks and outlinks of the connected websites to check the role of the FLG core websites in this network. The results are shown in Fig. 6. EpochTimes.com is mainly responsible for the linkage of human rights and democracy websites. Clearwisdom.net is located in the middle of the network and connects other major mixture websites. FalunDafa.org focuses on FLG cultivation and links local FLG practitioners' websites.

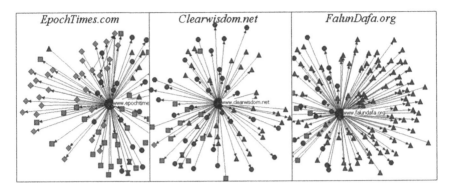

Fig. 6 Roles of FLG core websites in FLG network

4.2 Web Content Analysis

Writeprints illustrate the characteristics of words that authors frequently use to express their opinions or ideas. Figure 7 shows the Writeprint of Hongzhi Li. His discussion revolves around the teachings of Falun Dafa (法輪大法) and has neither significant temporal variation nor concentration on sub-topics as shown in Fig. 7(a). However, in his bag of words in Fig. 7(c), the word "evil" is used frequently.

The Writeprints of the editors of Clearwisdom.net, as shown in Fig. 8(a) and (b), had three significantly deviated areas which represented the topics of Dafa rectification (正法), righteous thoughts (發正念), and persecution of FLG practitioners (學員迫害真相). Their writings consistently revolved around these three major topics between 2000 and 2006. Comparing the Writeprints of Hongzhi Li and the editors allowed us to see their roles in the FLG movement. Hongzhi Li's articles focused on the central concepts of FLG cultivation but provide some hints of his political attitude (e.g., against evil). The editors' articles, on the other hand, provided their interpretations of Hongzhi Li's teaching.

4.3 Forum Content Analysis

4.3.1 Thread Topics

In the 740 threads collected from the forum "Falun Dafa Universal," ten main topics were identified based on the content of the threads: persecution accounts (學員迫害真相), FLG success story sharing (修煉心得分享), FLG ideology (法輪功哲學), FLG articles (法輪功書籍文獻), anti-Chinese Communist Party (anti-CCP, 反對中國共產黨), life philosophy (生活哲學), mysterious phenomena (宇宙科學與神秘現象), social issues (社會議題), health issues

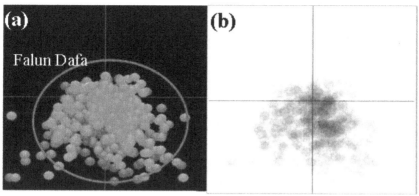

(c) **Key Word N-Gram and Ink Blot Features**

No.	Description	Usage	Mean	E
0	DAFA	0.868	0.037	0
1	FA	0.863	0.036	
2	PEOPLE	0.732	0.025	0
3	BEINGS	0.663	0.017	
4	THINGS	0.568	0.015	0
5	DISCIPLES	0.555	0.021	0
6	HUMAN	0.47	0.013	
7	EVIL	0.397	0.018	0
8	RECTIFICAT...	0.35	0.013	
9	COSMOS	0.263	0.0070	
10	CULTIVATION	0.26	0.01	0
11	TIME	0.252	0.011	

No.	Description	Usage	Mean
0	THE FA	0.348	0.014
1	OF THE	0.328	0.013
2	IN THE	0.323	0.012
3	DAFA DISCIPLES	0.299	0.011
4	FA RECTIFICA	0.223	0.0080
5	THE EVIL	0.158	0.0070
6	AND THE	0.133	0.0050
7	THE COSMOS	0.118	0.0030
8	TO THE	0.107	0.0050
9	THE OLD	0.105	0.0030
10	TO DO	0.1	0.0030
11	SENTIENT BEI...	0.093	0.0030

No.	Description	Weight	Blot
0	RIGHTEOUS	63.61	
0	HISTORY	25.81	
0	BEINGS	13.87	
0	DON	8.4	
0	GODS	5.52	
0	STUDENTS	3.94	
0	MASTER	3.09	
0	THINGS	2.7	
0	SITUATION	2.62	
0	MATTER	2.76	
0	ORDINARY	3.06	
0	HUMAN	3.47	
0	SOCIETY	3.97	
0	EVIL	3.47	

No.	Description	Usage
11	OF THE COSMOS	0.029
12	SENTIENT BEINGS	0.0
13	A DAFA DISCIPLE	0.027
14	VALIDATING THE FA	0.027
15	THE EVIL	0.0
16	THE HUMAN WORLD	0.024
17	IN OTHER WORDS	0.024
18	OTHER WORDS	0.0
19	BEINGS IN THE	0.022
20	NO MATTER HOW	0.022
21	DAFA DISCIPLES ARE	0.022
22	OLD FORCES	0.0

Fig. 7 Writeprint of Hongzhi Li

(健康議題), and general messages (網站管理訊息). The ten topics and their descriptions are listed in Table 6.

The distribution of threads over these ten topics is summarized in Table 7. Although life philosophy has the highest number of threads, anti-CCP has the highest average reply rate. Major discussions in this forum are often about anti-CCP topics.

Figure 9 displays the threads based on their similarity. We can see that the persecution accounts and anti-CCP are aligned and to some degree mixed together on the upper parts of the circle. Such a mixture

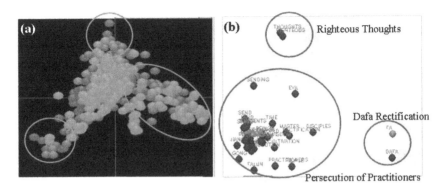

Fig. 8 Writeprint of the editors of Clearwisdom.net

Table 6 Ten main topics of forum threads

Main Topic	Description
Persecution accounts (學員迫害真相)	Detailed description of torture process and methods
FLG success story sharing (修煉心得分享)	Describe how a practitioner benefits from FLG
FLG ideology (法輪功哲學)	Share personal beliefs about Dafa rectification, cultivation, and righteous thoughts
FLG articles (法輪功書籍文獻)	Include articles and books from Hongzhi Li and FLG
Anti-CCP (反對中國共產黨)	Criticize CCP for organ harvest, human rights, and religious freedom
Life philosophy (生活哲學)	Share inspired life stories and words of wisdom
Mysterious phenomena (宇宙科學與神秘現象)	Distribute articles about the origin of the cosmos and unexplainable phenomena
Social issues (社會議題)	Discuss social issues, such as the role of news press and impact of violent video games
Health issues (健康議題)	Distribute health-related news and healthy recipes
General message (網站管理訊息)	Messages about the forum management and arguments

Table 7 Distribution of threads over ten topics

Main topic	Threads	Messages	Reply rate
Persecution accounts	100	214	1.14
FLG success story sharing	10	16	0.6
FLG ideologies	112	256	1.28
FLG articles	29	29	0
Anti-CCP	112	336	2
Life philosophy	166	255	0.54
Mysterious phenomena	87	107	0.23
Social issues	16	39	1.44
Health issues	86	105	0.22
General message	22	42	0.9

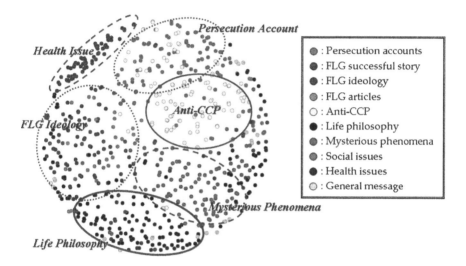

Fig. 9 Display of threads according to their similarity

is due to the high usage of the same key phrases, such as organ harvest (器官活摘). and labor camp (集中營). In the lower part, Falun Gong ideology is closely aligned with life philosophy and mysterious phenomena. From the relative positions of these three topics, we may infer that Falun Gong is similar to a religion, which not only teaches a certain life philosophy but also explains the origin of life.

From the display of threads, we can further see to which topics an author primarily contributes. Figure 10 shows the distribution of threads of the top two active authors in the forum. The author "Sujcs888" focused on Falun Gong ideology, life philosophy, and mysterious phenomena; while the author "LoveTender" targeted the topics of persecution and anti-CCP.

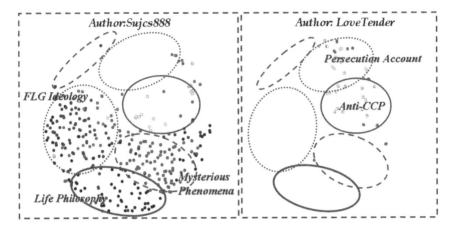

Fig. 10 Distribution of topics of the top two active authors in the forum

4.3.2 Author Interaction

In order to see which topics have shown more intense discussion among authors, we measured the interaction networks of the top five topics above with average degree and clustering coefficient. The average degree shows the overall activity or inter-action density of authors in a network. The clustering coefficient reflects clusters, which can indicate cliques or groups [35]. The results of these two measures are summarized in Table 8. The most intense interaction occurred in the discussion of FLG ideology and anti-CCP. However, compared to FLG ideology, anti-CCP had lower average degree but a much higher clustering coefficient. This implies that those authors were more likely to create discussion groups around this topic.

Table 8 Degree and clustering coefficient of five main topics

Main topic	Degree	Clustering coefficient
FLG ideology	2.400	0.086
Life philosophy	2.080	0.093
Mysterious phenomena	1.00	0.000
Persecution	1.733	0.000
Anti-CCP	2.261	0.128

Figure 11 shows the interaction networks relating to FLG ideology and the anti-CCP topic. Authors discussing FLG ideology seemed to follow the ideas and preaching of master practitioners (the node in the center of the network). In anti-CCP discussions, authors were more likely to share their opinions and interact with each other freely (as shown in small clusters of interactions).

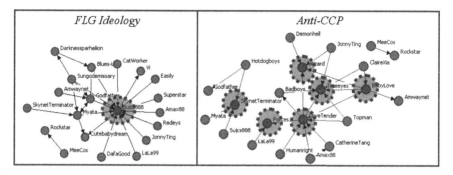

Fig. 11 Interaction network of authors in FLG ideology and anti-CCP

5 Conclusions

In this study, we took a cyber-archaeology approach and performed three separate analyses on the cyber-artifacts, including link, web content, and forum content analyses, to investigate the framing of FLG's identity in social movements. First, it is not surprising to see that FLG's websites closely link to two types of SMOs: Chinese democracy and human rights. This affiliation can be explained with the conflicts between FLG and the Chinese government: the official ban in 1999 and subsequent persecution of its practitioners. As Rahn [30] pointed out, those persecution accounts gave FLG a justified identity as a human rights movement. Second, by taking the cyber-archaeology perspective, we can see how FLG "strategically" deploys its cyber-artifacts forming its inclusive identity on the Internet. Not only are its four core websites located in the center of the FLG network, but they also bridge two seemingly unrelated groups of websites for different roles: activist and Qi-Gong groups. Each of the core websites plays a specific role in joining websites of different attributes. For example, most of human rights and democracies in the FLG network are connected via EpochTimes.com. In addition, we find a trace of its religious role in its forum: it tries to explain the origin and meanings of life. From topic clustering, we see how FLG includes and organizes several different concepts on a continuum: from FLG ideology to life philosophy and mysterious phenomena, and from mysterious phenomena to anti-CCP and persecution by conceptualizing the CCP as the "Evil." By deploying its cyber-artifacts, FLG smoothly connects different ideologies and establishes its inclusive role as a Qi-Gong, religious, and activist group.

As a religious group, Hongzhi Li without question is the spiritual leader of FLG. The Writeprint revealed a structural pattern resembling a religious hierarchy in the writings of Hongzhi Li and of the editors: Hongzhi Li's centered around the ideological teaching of Falun Dafa while the editors' posted specific programs outlining Li's teaching for the practitioners to follow. In the FLG forum, we also found that the authors exhibited different interaction patterns in the discussions of FLG ideology and of anti-CCP. They followed the monologue pattern of master-preaching in the FLG ideology but had more interaction in the anti-CCP discussion. Does the structural difference between a religious and an activist organization cause tension in FLG's cooperation with other SMOs? While we don't find any evidence from analyzing its cyber-artifacts to support this, the human rights torch relay, mainly hosted by CIPFG.org (Coalition to Investigate the Persecution of Falun Gong in China) between 2007 and 2008, revealed potential coordination problems between FLG and its allies. The relay was widely conceived as an FLG campaign against the Beijing 2008 Olympic Games and used as its slogan, "The Olympics and crimes against humanity cannot coexist in China." However, Hongzhi Li posted a message on April 4, 2008, explaining that the human rights torch relay is for everyday people and "so it is not Dafa disciples that this event is for." It is interpreted as a call for FLG practitioners to focus on FLG persecution and truth-clarification work rather than involvement in human rights activities. How the FLG is going to reframe its identity in social movements will require careful monitoring and future studies.

Acknowledgments Funding for this research was provided by NSF, "CRI: Developing a Dark Web Collection and Infrastructure for Computational and Social Sciences," 2007–2010.

References

1. Abbasi A, Chen H (2005) Applying Authorship Analysis to Extremist-Group Web Forum Messages. IEEE Intelligent Systems, 20: 1541–1672
2. Abbasi A, Chen H (2006) Visualizing Authorship for Identification. In: Intelligence and Security Informatics. San Diego, pp. 60–71
3. Ackland R, O'Neil M, Bimber B, Gibson RK, Ward S (2006) New Methods for Studying Online Environmental-Activist Networks. In: 26th International Sunbelt Social Network Conference, Vancouver
4. Almind TC, Ingwersen P (1997) Informetric Analyses on the World Wide Web: Methodological Approaches to Webometrics. Journal of Documentation 53: 404–426
5. Bell MR, Boas TC (2003) Falun Gong and the Internet: Evangelism, Community, and Struggle for Survival. Nova Religio 6: 277–293
6. Bennett W (2003) Communicating Global Activism. Information, Communication and Society 6: 143–168
7. Bennett WL (2005) Social Movements Beyond Borders: Understanding Two Eras of Transnational Activism. In: Porta Dd and Tarrow S (eds) Transnational Protest and Global Activism. Rowman & Littlefield Publishers, Inc., New York, pp. 203–226
8. Björneborn L, Ingwersen P (2001) Perspective of Webometrics. Scientometrics 50: 65–82
9. Björneborn L, Ingwersen P (2004) Toward a Basic Framework for Webometrics. Journal of the American Society for Information Science and Technology 55: 1216–1227
10. Clark JD, Themudo NS (2006) Linking the Web and the Street: Internet-based Dotcauses and the Anti-Globalization Movement. World Development 34: 50–74
11. Cohen R, Rai SM (2000) Global Social Movements. The Athlone Press, New Brunswick
12. Freeman LC (1978/79) Centrality in Social Networks: Conceptual Clarification. Social Networks 1: 215–239
13. Garrido M, Halavais A (2003) Mapping Networks of Support for the Zapatista Movement: Applying Social-Network Analysis to Study Contemporary Social Movements. In: McCaughey M, Ayers MD (eds) Cyberactivism: Online Activism in Theory and Practice. Routledge, New York

14. Gerlach LP (2001) The Structure of Social Movements: Environmental Activism and its Opponents. In: Arquilla J, Ronfeldt DF (eds) Networks and Netwars: The Future of Terror, Crime, and Militancy. Rand, Santa Monica, pp. 289–309
15. Jones Q (1997) Virtual-Communities, Virtual Settlements & Cyber-Archaeology: A Theoretical Outline. Journal of Computer-Mediated Communication 3
16. Jones Q, Rafaeli S (2000) What Do Virtual "Tells" Tell? Placing Cybersociety Research into a Hierarchy of Social Explanation. In: Hawaii International Conference on System Sciences, Hawaii
17. Keck ME, Sikkink K (1998) Activists Beyond Borders: Advocacy Networks in International Politics. Cornell University Press, New York
18. Klovdahl AS, Potterat JJ, Woodhouse DE, Muth JB, Muth SQ, Darrow WW (1994) Social networks and Infectious Disease: the Colorado Springs Study. Social Science & Medicine 38: 79–88
19. Langman L (2005) From Virtual Public Spheres to Global Justice: A Critical Theory of Internetworked Social Movements. Sociological Theory 23: 42–74
20. Larana E, Johnston H, Gusfield JR (1994) New Social Movements: From Ideology to Identity. Temple University Press, Philadelphia
21. Lu Y (2005) Entrepreneurial Logics and the Evolution of Falun Gong. Journal for the Scientific Study of Religion 44: 173–185
22. McAdam D, McCarthy J, Zald M (1999) Comparative Perspectives on Social Movements, Cambridge University Press, Cambridge
23. McCaughey M, Ayers MD (2003) Cyberactivism: Online Activism in Theory and Practice. Routledge, New York
24. Morris AD, Mueller CM (1992) Frontiers in Social Movement Theory. Yale University Press, New Haven
25. Myers DJ (1994) Communication Technology and Social Movements: Contributions of Computer Networks to Activism. Social Science Computer Review 12: 250–260
26. Ong TH, Chen H (1999) Updateable PAT-Tree Approach to Chinese Key Phrase Extraction using Mutual Information: A Linguistic Foundation for Knowledge Management. In: Proceedings of the Second Asian Digital Library Conference, Taipei
27. Paccagnella L (1997) Getting the Seats of Your Pants Dirty: Strategies for Ethnographic Research on Virtual Communities. Journal of Computer-Mediated Communication 3
28. Penny B (2003) The Life and Times of Li Hongzhi: Falun Gong and Religious Biography. The China Quarterly 175: 643–661
29. Porta Dd, Diani M (1999) Social Movements: An Introduction. Blackwell Publishers Ltd, Malden
30. Rahn P (2002) The Chemistry of a Conflict: The Chinese Government and the Falun Gong. Terrorism and Political Violence 14: 41–65
31. Rosenthal N, Fingrutd M, Ethier M, Karant R, McDonald D (1985) Social Movements and Network Analysis: A Case Study of Nineteenth-Century Women's Reform in New York State. The American Journal of Sociology 90: 1022–1054
32. Scott J (2000) Social Network Analysis: A Handbook, 2nd edn. Sage, London
33. Tesh SN (2002) The Internet and the Grass Roots. Organization & Environment 15: 336
34. Tong J (2002) An Organizational Analysis of the Falun Gong: Structure, Communications, Financing. The China Quarterly 171: 636–660
35. Wasserman S, Faust K (1994) Social Network Analysis: Methods and Applications. Cambridge University Press, New York

Patterns of Word Use for Deception in Testimony

David B. Skillicorn and Ayron Little

Abstract Patterns of deception in word use are well-known, but interrogative settings, particularly court testimony, impose additional constraints that appear to alter these patterns. We suggest altered patterns applicable in interrogative settings, and validate them using testimony to a Canadian commission. Predictive accuracy rates above 80% are achieved. Difficulties remain, however, because other work suggests that emotional intensity also impacts the patterns associated with deception, so these results may not extend to criminal trials.

Keywords Discourse · Textual analysis · Lying · Word use

1 Introduction

Detecting when other humans were being deceptive has been an important problem for as long as humans have lived in social groups. From an evolutionary and social perspective, we might expect that a kind of balance, in effect a stalemate, exists between the ability to be deceptive successfully and the ability to detect deception. An improvement on either side would confer a substantial advantage that would tend to spread, and yet we do not see widespread success on either side.

On the other hand, being deceptive, inventing experiences that never happened, must surely require substantial mental resources. We might expect to see some signature of this extra effort whenever deception was taking place. In what follows, we will suggest an explanation for this apparent contradiction.

There is a long tradition of looking for deception in law-enforcement and forensic settings. Much of this has relied on intuition, often contradictory. For example, making too little eye contact is often regarded as a sign of being deceptive – but

David B. Skillicorn, Ayron Little
School of Computing, Queen's University, Kingston, Canada
e-mail: skill@cs.queensu.ca

C.C. Yang et al. (eds.), *Security Informatics*, Annals of Information Systems 9,
DOI 10.1007/978-1-4419-1325-8_2, © Springer Science+Business Media, LLC 2010

so is making too much eye contact. In tests where law-enforcement personnel were asked whether or not certain witnesses were being deceptive, the success rate was very little higher than chance [10].

Nevertheless, empirical investigation of reliable signs of deception has been carried out, and some usable results are known. In this chapter, we start from an empirical model of deception in text (written directly or transcribed from speech) developed by Pennebaker's group [9]. This model applies to freeform text, where the author is able to express himself or herself in any way. The model is based on changes in frequencies of 88 significant words, primarily pronouns, conjunctions, and common verbs and adjectives. Such "little" words are a tiny percentage of available words, but make up around half of words used.

Most law-enforcement and forensic settings, however, are interrogative, so that the content generated by a potentially deceptive subject is constrained by the form of the questions being asked, and the need to appear responsive. Furthermore, in some such settings, the kinds of questions that will be asked are partially predictable, so someone attempting to be deceptive has both the motivation and the ability to rehearse possible responses. These constraints require different ways of forming utterances and, in particular, may alter the form of deceptive responses. We present a modification to the Pennebaker model for court testimony, and validate it in the context of a Canadian judicial inquiry. Our results show better than 80% agreement with media judgments of which witnesses might have been motivated to be deceptive.

2 Related Work

Humans communicate using a number of modalities, visual, auditory, and content-filled, and it is possible to consider detecting deception in all of them. For example, analysis of video can allow microexpressions to be detected [5], some of which indicate deception. In the auditory channel, deception may be signaled by a quavering voice or changes in tempo, because the mental processes associated with deception have physiological consequences [8]. In fact, a key fact in considering deception is that the subconscious awareness of those being deceptive leaks out into their actions and words in ways that they are helpless to prevent. If these subconsciously generated signals can be detected, they are highly reliable because they are so difficult to conceal or manipulate.

Techniques for detecting deception generally make two assumptions. The first is that having a motivation to deceive increases the level of anxiety in the deceiver, and this in turn leads to increases in the level of errors in language formation and the like [1]. The second is that creating deceptive content is cognitively difficult. Because it requires doing something (creating discourse) consciously that is normally done subconsciously, attempts to be deceptive typically lead to overcontrol and purpose tremor which, in turn, can create detectable anomalies. Also the effort required to create content may lead to increased abstraction and to omission of otherwise-typical content because there are not enough mental resources to present a fully rounded discourse. A good summary of work is detecting deception in all of these modalities is [4].

We will consider only the detection of deception from text (and so also from speech by transcription) in what follows. Using video and audio modalities has some drawbacks. The data capture is relatively expensive, because it may require full-motion video and special-purpose hardwares such as eye trackers and temperature and blood pressure detectors. Detection can require significant training, and there are often issues related to coding and consistency. For some technologies, those who wish to be deceptive may be able to learn how to reduce the signatures being looked for using relaxation or hypnotic techniques. In contrast, text is cheap to gather and, because language generation is so subconscious, it is extremely difficult to hide the signals of deception, even with knowledge of what they are.

2.1 The Pennebaker Model of Deception

The model of deception developed by Newman et al. [9] is based on changes in the frequencies of a set of 88 words. The model was developed empirically – individuals were asked to write short arguments in support of positions they either did or did not hold. Characteristic differences in the rates of usage of the words in the model emerged. Model building was done initially using student subjects, but has been extended to large numbers of subjects from many different settings, and the relevant words have remained consistent. The significant words fall into four categories:

1. First-person singular pronouns ("I," "me"). The rate at which these are used decreases when someone is being deceptive, perhaps as a way of distancing themselves from the fabricated content.
2. Exclusive words ("but," "or"), words that introduce refinements of the content. The rate at which these are used decreases, perhaps because of the cognitive difficulty of producing refinements for events and actions that did not actually occur.
3. Negative-emotion words ("hate"). The rate at which these are used increases, perhaps because a self-awareness that deception is socially disapproved leaks into the language used.
4. Motion verbs ("go," "move"). The rate at which these are used increases, perhaps as a way of keeping the story moving and preventing the listener/reader from thinking more deeply about the content.

The full set of words in each of the word classes is given in Table 1.

The form of the model shows how it can be that deception is visible in content, but we, as humans, are unable to detect it. Deception causes changes in the frequencies of use of these words, but human listeners and readers are not equipped to count the frequencies of 88 different words, and simultaneously judge whether they are being used at typical, higher, or lower rates than usual. However, given a text, software can easily accumulate the relevant frequencies and give each document a deceptiveness score.

Table 1 Words of the Pennebaker deception model

Word class	Words
First-person singular pronouns	I, me, my, mine, myself, I'd, I'll, I'm, I've
Exclusive words	but, except, without, although, besides, however, nor, or, rather, unless, whereas
Negative-emotion words	hate, anger, enemy, despise, dislike, abandon, afraid, agony, anguish, bastard, bitch, boring, crazy, dumb, disappointed, disappointing, f-word, suspicious, stressed, sorry, jerk, tragedy, weak, worthless, ignorant, inadequate, inferior, jerked, lie, lied, lies, lonely, loss, terrible, hated, hates, greed, fear, devil, lame, vain, wicked
Motion verbs	walk, move, go, carry, run, lead, going, taking, action, arrive, arrives, arrived, bringing, driven, carrying, fled, flew, follow, followed, look, take, moved, goes, drive

It is clear that, if a decrease in word frequency is significant, there must be some baseline against which that decrease can be observed. This means that the model can only be applied to a set of texts from, in some sense, the same domain, so that the norms of usage determine the base rates of word usage, against which increases and decreases can be measured. For example, first-person pronouns are rarely used in business writing, so such writing would seem relatively more deceptive when compared to other, more informal, texts.

The Pennebaker model treats each word as an equally important signal of deception. However, it may be that, in some settings, very small changes in the frequency of one word are important signals of deception, while large changes in another are not. It is therefore also important to consider correlation among different words, in each particular set of texts. We have extended the Pennebaker model using singular value decomposition (SVD) as a way to address this issue. This can be thought of as a kind of data set-specific factor analysis.

Another investigation of open-ended written texts that is relevant is the work of Adams [1]. She examined statements by victims and perpetrators of crimes for which the actual facts were known. She discovered support for the following associations:

1. Equivocation (the use of words such as "maybe" and "perhaps," as well as decreased rates of first-person pronouns) is positively correlated with deception.
2. Negation (the use of words such as not, which allow statements that avoid describing what did happen, in other words a kind of omission) is positively correlated with deception.
3. The length of the prologue (that part of the statement before the description of actual events begins) is positively associated with deception.
4. The inclusion of unique sensory details (particular sounds, sensations, smells, and so on) is positively correlated with veracity.

She also showed that emotion in the concluding section of a statement had a weak association with veracity, and that any association between the presence of quoted discourse and veracity is very weak. (Such an association has been posited by others.) These associations are broadly consistent with the Pennebaker model.

2.2 Including Correlation in the Pennebaker Model

The counting of frequencies in a set of documents produces a matrix, say A, with one row for each text, and one column for each model word from the Pennebaker model; the entries are the frequencies of each word in each text, usually normalized in some way. Each row of this matrix can be considered as a point in a space spanned by axes corresponding to each of the words. Performing an SVD on this matrix produces new coordinates relative to a new set of axes that capture the important correlated variation in the entire set of words.

The singular value decomposition [6] of matrix A expresses it as the product of three matrices:

$$A = U S V'$$

where the rows of V' represent a new set of axes, U represents a new set of coordinates for each row of A relative to these new coordinates, and S is a diagonal matrix with non-increasing entries. The new axes have the property that the first new axis is oriented in the direction of maximal variation among the rows of A; the second axis is oriented in the direction of remaining (orthogonal, uncorrelated) maximal variation, and so on. The entries of the diagonal of S estimate how much of the variation among the rows is captured in each dimension. A useful property of the SVD, therefore, is that the right-hand side can be truncated. This allows the data to be faithfully represented using many fewer dimensions than are required in the original representation of the data, and the amount of information lost by doing so can be estimated from the diagonal elements of S that are discarded by the truncation.

The position of each text in this new coordinate space can be interpreted as a deceptiveness score in a way that takes into account correlations in the way words are used in the domain being considered. For large sets of documents this makes a significant difference [7].

2.3 Interrogation

There are two forms of interrogative discourse in which deception is important, but which are substantially different in their properties. In many law-enforcement settings, questioning may be wide-ranging and repetitive, and there are chances to follow up and repeat questioning about any issue. These settings are not public and usually involve only one subject being questioned at a time. In contrast, the questioning in court testimony is highly constrained in form and time, and takes place publicly, but the lines of questioning can often be better anticipated.

Pennebaker (J.W. Pennebaker and D.A. Huddle, 2007, personal communication) has also extended his model to consider court testimony in situations where an individual was convicted and subsequently exonerated based on DNA evidence, so that ground truth is known. His results show that changes in word usage depend on the emotional level of the testimony. In emotionally charged testimony, deception is characterized by reductions in the use of first-person singular pronouns, but in less emotionally charged testimony, deception is characterized by increases in first-person singular pronouns. Unfortunately, juries are prone to interpret increases in first-person pronoun use as signals of deception. An innocent defendant is most likely to give emotionally charged but honest testimony, characterized by high first-person singular pronoun use, and so is likely to be disbelieved.

3 A New Model of Deception

We consider court testimony in situations that are not emotionally charged and suggest a new model that the signal of deception in such settings is:

1. Increased use of first-person pronouns. There are several reasons why this is plausible. In court, many questions are of the form "Did you ...?" that necessarily invite responses of the form "I did," or "I did not." Responses using first-person singular pronouns also make it possible to answer a question and provide justification at the same time. It also seems plausible that deceptive respondents might use the pronoun "we" extensively, in response to questions of the form "Did you ..." and we explore this issue later. It has also been observed previously [2] that the lower-status person in an interaction tends to use first-person singular pronouns at higher rates, and the trappings of a courtroom tend to create the impression that a witness is of lower status than counsel and the judge.
2. Increased use of exclusive words. In freeform statements, it is difficult to be deceptive by using exclusive words at high rates because they increase the required complexity. However, in a testimony situation, it is likely that a witness motivated to be deceptive will have given some thought to likely lines of questioning, and to potential ways to answer them. This element of rehearsal reduces the cognitive load at the moment of answering the question. Exclusive words also provide a mechanism for justification, for example, "Did you ..." answered by "Yes, I did but ...." Exclusive words also allow hypothetical possibilities to be introduced into answers, providing a way to "muddy the waters."
3. Increased use of negative-emotion words, perhaps at lower levels than in freeform text. In a courtroom situation, rehearsal may have lowered the emotional tone, because the questions and answers will have been repeated several

times beforehand and witnesses will, of course, attempt to appear calm and sure of themselves as a general strategy for avoiding the appearance of guilt.
4. Increased use of motion verbs for the same reason as before, to keep the story flowing so that the hearers do not have time to consider possible flaws.

In this revised deception model, it is *increases* in frequency that signal deception in all four word classes, so the issue of using a set of documents that are all of the same form is less directly important. However, it is still the case that correlations among words are important, and that some words are more significant than others, so we continue to use an SVD as an important part of the analysis task.

We validate this model using a data set of text extracted from the testimony to the Gomery Commission [3], a judicial investigation instigated by the Canadian government after the so-called Sponsorship Scandal. The Canadian government had disbursed large sums of money to be spent convincing the province of Quebec of the advantages of remaining in Canada. A large fraction of this money was apparently diverted at various stages of the process to other purposes including the enrichment of some of the participants. The Gomery Commission was set up, without a criminal mandate, to investigate what had happened.

4 Tuning the Model

Before we can apply this revised model to testimony data, we must consider a number of preparatory steps. The basic algorithm is this:

1. Count the frequencies of each of the model words in each text, creating a matrix with a row for each text and a column for each model word.
2. Perform a singular value decomposition on the resulting matrix, producing a matrix with a row for each document, and values that reflect the positions of each text with respect to a new set of axes.
3. Interpreting these positions as deceptiveness scores.

There are design decisions to be made for each of these stages, and we now consider them.

4.1 Normalization

In general, we do not want long documents to exert an undue effect on the model simply because the frequencies of model words in long documents will tend to be larger than the frequencies in short documents. However, normalization is not trivial because, roughly speaking, words are of two kinds. Some words occur regularly in text – for example, the word "the" occurs about every tenth word in English –

so dividing their frequencies by the length of the document accurately captures variations in their use. Other words, however, occur much less regularly, and it may be their absolute occurrences, independent of the length of a document, that are most significant. We have explored this issue extensively, but there is no straightforward solution and so, for this experiment, we normalize frequencies by dividing them by the length of each text in words.

The symmetry of the SVD with respect to both texts and words allows us also to visualize a space occupied by the words and assess their relative importances in the model. (This is one of the advantages of using SVD rather than Principal Component Analysis.) Doing so shows very clearly that some words are very much more significant than others.

We make use of this information to simplify the model. Instead of counting the frequencies of each of the 88 words individually, we divide them into six classes and count the frequencies within each class instead. The classes are: the first-person singular pronouns, "but," "or," the remaining exclusive words, the negative-emotion words, and the motion verbs. In other words, the frequency of "but" is treated as of equivalent importance to the frequencies of all of the negative-emotion words. Figure 1 shows a visualization of all of the words. It is clear that the words within each of the four classes do not form clusters; rather, some words have quite different effects to others. Words from each class that are important and different from the other members of their class are retained as explicit factors; the remaining words, with similar effects to each other, are retained collectively.

Figure 2 shows the visualization of the words in six classes. It is clear that first-person singular pronouns, "but," "or," and motion verbs play much more significant

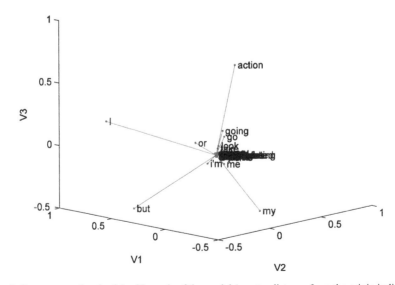

Fig. 1 Importance of each of the 88 words of the model (greater distance from the origin indicates greater importance)

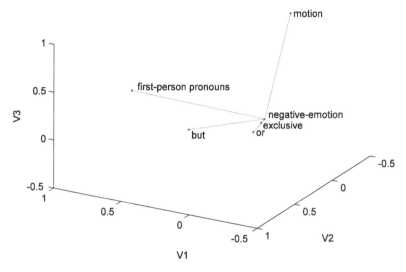

Fig. 2 Importance of each of the six word classes

roles than negative-emotion words and the other exclusive words, and that each of these classes plays a different role to the others.

Some of the testimony to the Commission was given in French. In this experiment, we have used the official translations to English. However, this introduces some uncertainty, since French allows a pronoun "on" that can play the role of "I," but also roles that require other pronouns in English. (The fact that many Francophone witnesses chose to testify in English is itself interesting.)

5 Experiment

The testimony to the Gomery Commission was downloaded and the word frequencies for the entire testimony of each participant were extracted. A total of 160 witnesses testified, so the resulting matrix has 160 rows and six columns. The frequencies in each row were normalized by the total number of words used by each witness, for the reasons described above.

The entries in each column were then normalized, by subtracting the column mean from each entry, and dividing by the column standard deviation (that is, converting to z-scores). The subtraction of the column mean ensures that the values in each column are centered at zero, a necessity for the SVD. The division by the standard deviation makes the magnitudes of the normalized frequencies of each word count approximately equal, compensating for their differing base frequencies in English.

If deception were a single-factor property, it would suffice to project the points corresponding to each individual's text onto the first axis of the transformed space, and interpret this as a deceptiveness score, relative to the set of texts under consideration. However, it is quickly obvious that deception is a more complex property, so it is helpful to look at plots in two or three dimensions, so that the relationships among the texts can be fully assessed.

It is also useful, however, to project the points corresponding to each text in a way that allows them to be given a single simple score. We do this by drawing a vector from the origin to the point (s_1, s_2, s_3), where these are the first three values of the diagonal of the S matrix. This line can be interpreted as a deceptiveness axis since its direction takes into account the relative importance of variation along the first three transformed axes. The points corresponding to each text can then be projected onto this line to give them each a "deceptiveness score." Since there is an ambiguity in the SVD, because any column of U and the corresponding row of V' can be multiplied by -1 without changing the result, care must be taken that the direction of this deceptiveness axis is correct. We do this by inserting an artificial point with the word frequency patterns of a deceptive text to provide the correct orientation for the line.

Because of the numerical properties of the SVD, it is possible to upweight a row of the matrix (that is, move it further from the origin) to see whether this has an effect on other rows (or indeed columns). This works because SVD takes full account of the correlations of every order between the rows and the columns. Unfortunately, upweighting texts of those with particularly strong reasons to be deceptive did not reveal anything interesting. This is probably because the participants in the scandal and associated fraud were not very tightly coupled – the frauds did not depend on high levels of collusion. Arguably, some of the participants regarded what they were doing as perquisites rather than fraud.

6 Results

The fundamental result of the analysis is shown in Fig. 3.

The positive end of the deceptiveness axis extends to the right (texts that are more deceptive than average), while the negative end extends to the left. The testimony of witnesses 1 and 2 included statements denying knowledge of what they had been paid to do, how they were paid, and who they worked for. It seems eminently plausible that this is highly deceptive, and that was certainly the opinion of Gomery. Witness 160, at the other end of the list, was not involved in the events and gave evidence of a purely factual kind, explaining how advertising contracts were typically awarded.

Projecting the points corresponding to individuals' testimony onto the deceptiveness axis clearly loses significant information about differences in patterns of word usage, but there does not seem to be particularly interesting structure in the clustering of the points. For example, witness 1 has a high deceptiveness score based heavily on the use of first-person singular pronouns, while witness 2 uses "but" and "or" heavily and also uses motion verbs at non-trivial levels.

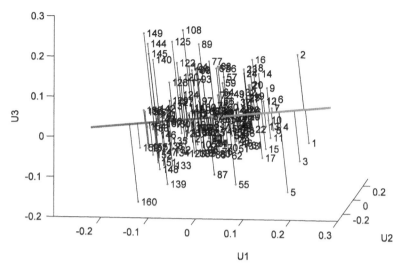

Fig. 3 Projection of witness speeches onto a "deceptiveness" axis that is oriented along the direction of maximal variation in the first three dimensions of the transformed space

To validate this deceptiveness score, we determined who might have been motivated to be deceptive based on:

- The assessments of the Gomery Commission itself. Individuals faced penalties ranging from referral for criminal prosecution to expulsion from the Liberal party.
- Assessments by the media of those who had probably been involved in some malfeasance.

This produced a list of 71 of 160 witnesses who had motivation to be deceptive. The following table gives the confusion matrix for prediction, assuming that the top 71 ranked texts are labeled as deceptive.

	Predicted deceptive	Predicted non-deceptive
Reason to be deceptive	56	15
No reason to be deceptive	15	74

This gives a prediction accuracy of 81.3%, with 8.1% of those presumptively deceptive predicted to be non-deceptive, and 9.4% of those presumptively non-deceptive predicted to be deceptive.

Because all of the word frequencies in the modified deception model increase with deception, the use of SVD, rather than simply summing the per-word scores, is less significant. However, it is still helpful to use correlation information obtained from the SVD. A simple summed-score model has a prediction accuracy of 80%.

An alternative would be to treat those texts with positive deceptiveness scores as deceptive and those with negative deceptiveness scores as non-deceptive. This produces the following confusion matrix.

	Positive deception score	Negative deception score
Reason to be deceptive	60	11
No reason to be deceptive	21	68

This gives a prediction accuracy of 80%. Moving the decision boundary increases the number of false positives and decreases the number of false negatives because there are 81 individuals with a positive deception score, but only 71 motivated to be deceptive.

To further validate the model, we included the texts of the lawyers. Although lawyers are conventionally regarded as chronically deceptive, in this setting they did not pay a fundamentally advocacy role. Thus they had little motivation to be deceptive in their questioning. Their language patterns should be substantially different from those of witnesses, although many of the content words will overlap, because they are asking questions and witnesses are answering them, and because there are professional conventions about the way in which questioning takes place in court, it is a highly constrained form of discourse. The plot of all of the texts, including those by the lawyers, can be seen in Fig. 4. The points forward and to the right correspond to witnesses and the points at the rear and upwards to the right to

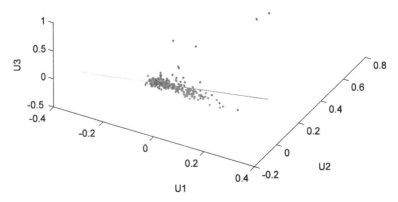

Fig. 4 Points corresponding to witness testimony (*magenta*) and lawyers' statements (*blue*). The content of lawyers' statements is both non-deceptive and qualitatively somewhat different from that of witnesses

lawyers. It is clear that almost all of the points corresponding to the lawyers' texts are at the non-deceptive end of the deceptiveness axis, and that their language patterns are somewhat different from those of witnesses. The "cloud" of lawyer points lies in a slightly different direction from that of the witnesses, with some of its extremal points very far from the other texts. The least deceptive lawyer text was that by the Clerk of the Court, whose speeches, and so text, were entirely formal.

It is notable that the pronoun "we" is not part of the deception model. This is intuitively surprising, since its use would seem to be an ideal way to diffuse responsibility and blame. For example, the response to "Did you break into the house" might be "We went in the direction of the house," providing a way for witnesses to distance themselves from particular actions, and also to involve others in them.

However, the pronoun "we" is used in ways that are completely orthogonal to the use of other words in the deception model. Increased use of "we" does not come at, for example, the expense of decreased use of "I." Instead some people habitually use both at high rates, some use one rather than the other, and some hardly use either. Hence, the rate of use of "we" provides no information about deceptiveness.

This can be seen in Fig. 5, which shows the importance of the word classes from the deception model, together with "we." The vector corresponding to "we" is more or less orthogonal to the vectors belonging to the other word classes. The vector is long, indicating that there are important variations between individuals in the way they use this pronoun, but these are uncorrelated with other words that signal deception.

Work using this deception model has so far been restricted to texts in English. Since the model is based on underlying assumptions about the psychology of deception, we can plausibly expect that it will work in approximately the same fashion in

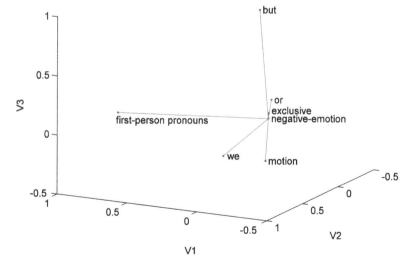

Fig. 5 Importance of word classes when the pronoun "we" is included. Its role is orthogonal to that of model words, so its use is uncorrelated to deception

other languages, although the precise set of words might vary. For example, as we have mentioned, the French pronoun *on* (analogous to the almost archaic English pronoun *one*) can be used in place of a first-person singular pronoun. This would obviously impact the model. Cultural differences may also be significant. The existing model appears to be partially driven by a psychological process connected to an underlying sense of shame in the person being deceptive. In some cultures, norms about the extent to which deceptiveness is considered shameful differ, which may also alter the model.

7 Conclusion

The Pennebaker model of deception in freeform text predicts changes in the frequency of use of four classes of words: increases for two classes and decreases for the other two classes. We suggest a variation of this model for detecting deception in non-emotional testimony, where the interrogative form, the power structure, and the possibility of predicting plausible questions and so rehearsing answers exist. In the revised model, deception is signaled by increases in all four classes of words.

We validate this model using testimony to the Gomery Commission, a Canadian investigation of widespread misuse of government funds, in which, unlike a typical criminal trial, many witnesses had some motivation to be deceptive. We achieve greater than 80% agreement between those predicted to have been deceptive and those who were judged, by the Commission and by the media, to have had a motivation to be deceptive.

The language structures reflecting deception in interrogative situations are still not well-understood because of the apparent role of rehearsal and emotional content. In criminal interrogation, rehearsal may not play much of a role, whereas in a trial setting rehearsal may be significant. This has not yet been explored in detail. In both situations, those with something to conceal may be more emotional, and this will change the way in which deception is signaled relative to those who are less emotional. In particular, in freeform text, interrogation, and emotional court testimony, deception appears to be signaled by decreasing first-person singular pronoun use, while in less emotional testimony, deception appears to be signaled by increasing first-person singular pronoun use. The need to judge how emotional a given interaction is before being able to use the model is, at present, an impediment to practical use.

References

1. S.H. Adams. *Communication Under Stress: Indicators of Veracity and Deception in Written Narratives*, Ph.D. Thesis, Virginia Polytechnic Institute, 2002.
2. C.K. Chung and J.W. Pennebaker. The psychological function of function words. In K. Fiedler, editor, *Frontiers in Social Psychology*. Psychology Press, New York, 343–359, 2007.

3. Commission of Inquiry into the Sponsorship Program and Advertising Activities. Gomery Transcripts. www.gomery.ca/en/transcripts/index.asp, 2004.
4. B.M. DePaulo, J.J. Lindsay, B.E. Malone, L. Muhlenbruck, K. Charlton, and H. Cooper. Cues to deception. *Psychology Bulletin*, 9:74–118, 2003.
5. P. Ekman. *Telling Lies: Clues to Deceit in the Marketplace, Marriage, and Politics*, 3rd edition. W.W. Norton, New York, 2002.
6. G.H. Golub and C.F. van Loan. *Matrix Computations*, 3rd edition. Johns Hopkins University Press, Baltimore, 1996.
7. S. Gupta. *Modelling Deception Detection in Text*. Master's thesis, School of Computing, Queen's University, 2007.
8. T.O. Meservy, M.L. Jensen, W.J. Kruse, J.K. Burgoon, and J.F. Nunamaker Jr. Automatic extraction of deceptive behavioral cues. In A. Silke, B. Ganor, and K. Joslin, editors, *Terrorism Informatics*. Springer, New York, 495–516, 2008.
9. M.L. Newman, J.W. Pennebaker, D.S. Berry, and J.M. Richards. Lying words: Predicting deception from linguistic style. *Personality and Social Psychology Bulletin,* 29:665–675, 2003.
10. A. Vrij. *Detecting Lies and Deceit: The Psychology of Lying and the Implications for Professional Practice*. John Wiley, Chichester; New York, 2002.

Information Integration for Terrorist or Criminal Social Networks

Christopher C. Yang and Xuning Tang

Abstract Social network analysis discovers knowledge embedded in the structure of social networks, which is useful for intelligence and law enforcement force in investigation. However, individual agency usually has part of the complete terrorist or criminal social network and therefore some crucial knowledge could not be extracted. Sharing information between different agencies will make such a social network analysis more effective, unfortunately the concern of privacy preservation usually prohibits the sharing of sensitive information. There is always a trade-off between the degree of privacy and the degree of utility in information sharing. Several approaches have been proposed to resolve such dilemma in sharing data from different relational tables. However, there is only limited amount of work on sharing social networks from different sources and yet trying to minimize the reduction on the degree of privacy. The work on privacy preservation of social network data relies on anonymity and perturbation. These techniques are developed for the purpose of data publishing, but ignore the utility of the published data on social network analysis and the integration of social networks from multiple sources. In this chapter, we propose a sub-graph generalization approach for information sharing and privacy preservation of terrorist or criminal social networks. The objectives are sharing the insensitive and generalized information to support social network analysis but preserving the privacy at the same time. Our experiment shows that such an approach is promising.

Keywords Social network · Information sharing · Privacy protection · Intelligence and security informatics

Christopher C. Yang, Xuning Tang
College of Information Science and Technology, Drexel University, Philadelphia, PA, USA
e-mail: chris.yang@drexel.edu

1 Introduction

Terrorist or criminal social networks have been proven to be useful for investigation, identifying suspects and gateways, and extracting communication patterns of terrorist or criminal organizations [5, 19, 20, 21]. An illustration of a terrorist social network as presented in [19] is shown in Fig. 1. However, an accurate result cannot be obtained if one only has a partial terrorist or criminal social network. Some important patterns or knowledge cannot be extracted in such case. In reality, no agency has a complete social network, but each agency only has information about some of the terrorists and their relationships. As a result, it is difficult to extract the timely and crucial knowledge for developing strategies in combating terrorism or crimes effectively unless information can be shared. From this perspective, it is ideal if all agencies are able to share their social networks and integrate them as a global social network. Unfortunately, the social network owned by individual agency may contain sensitive information. Such sensitive information is usually confidential due

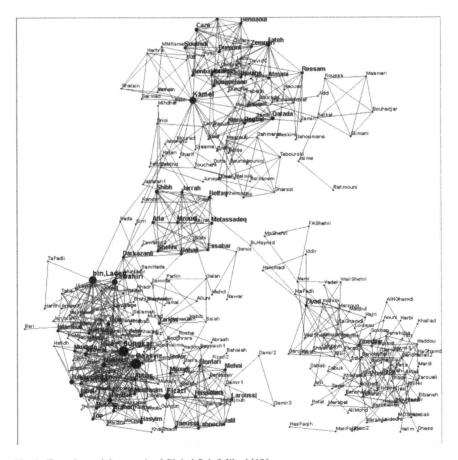

Fig. 1 Terrorist social network of Global Salafi Jihad [19]

to the privacy issue that causes the limitation on the degree of information sharing. Ideally, it is desired to share the crucial information in order to mine as much knowledge as if we have the complete information without violating the privacy. If we increase the degree of information sharing, the utility of the shared information increases and more knowledge can be extracted but the degree of privacy decreases. On the contrary, if we decrease the degree of information sharing, the degree of privacy increases but the utility of the shared information decreases. The challenge is how to develop a mechanism so that the information with high utility can be identified and shared with the minimum impact on privacy.

Information comes from data sources at different security levels and different agencies, for example, Federal Bureau of Investigation (FBI), Central Intelligence Agency (CIA), air force, army, navy, local, state and federal law enforcement agencies, and other private agencies. If information sharing is not available among these agencies, a complete picture of the terrorist community or crime organization cannot be obtained. Investigation will be delayed and the consequence can be severe. Indeed homeland security policy contends that, on the local level, information gathering, coordination with state and federal agencies, infrastructure protection, and enhanced development of police–community relationships will facilitate prevention and aid response to potential terrorist attacks [4, 6, 7, 14].

Thuraisingham [16] defined assured information sharing as information sharing between organizations but at the same time enforcing security and integrity policies so that the data is integrated and mined to extract nuggets. Thuraisingham described a coalition, in which members may join and leave the coalitions in accordance with the policies and procedures [16]. Members or partners conduct coalition data sharing in a dynamic environment. Baird et al. [3] first discussed several aspects of coalition data sharing in the Markle report. Thuraisingham [16] has further discussed these aspects including confidentiality, privacy, trust, integrity, dissemination, and others.

In this work, we focus on the social network information sharing and privacy preservation. We develop a sub-graph generalization method to share the only necessary information so that privacy can be protected. As a result, mining can be conducted to discover further knowledge to support social network analysis.

A social network is a graph, $G = (V, E)$, where V is a set of nodes representing persons and E is a set of edges ($V \times V$) representing the relationships between the corresponding persons.

In our research problem, the goal is sharing the information between social networks from different sources in order to conduct more accurate social network analysis for intelligence investigation. Without information sharing, we only have partial information of the complete social network that can be merged from these social networks. However, due to the privacy issue, releasing the information of the nodes and edges of one social network from one source to another is not permissible.

Example 1: Given a social network $G_1 = (V_1, E_1)$ and $G_2 = (V_2, E_2)$, we can merge G_1 and G_2 to G if information sharing is possible.

$V_1 = (A,B,C,D,E,F,G,H)$

Fig. 2 G_1, G_2, and the G_1: G_2:
integrated graph G

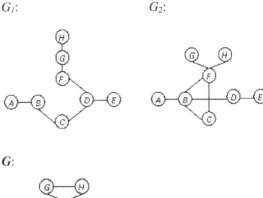

G:

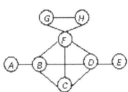

Table 1 Adjacency matrix of G

A	0	1	0	0	0	0	0	0
B	1	0	1	1	0	1	0	0
C	0	1	0	1	0	1	0	0
D	0	1	1	0	1	1	0	0
E	0	0	0	1	0	0	0	0
F	0	1	1	1	0	0	1	1
G	0	0	0	0	0	1	0	1
H	0	0	0	0	0	1	1	0

$V_2 = (A,B,C,D,E,F,G,H)$
$E_1 = ((A,B), (B,C), (C,D), (D,E), (D,F), (E,F), (F,G), (G,H))$
$E_2 = ((A,B), (B,C), (B,D), (B,F), (C,F), (D,E), (F,G), (F,H))$

Figure 2 shows G_1, G_2, and G. Table 1 shows the adjacency matrix of G.
Some common social network analysis techniques are

$$\text{degree centrality } (u) = \frac{\text{degree of } u}{(n-1)} \tag{1}$$

$$\text{closeness centrality } (u) = \frac{n-1}{\sum_{v=1}^{n} d(u,v)} \tag{2}$$

where $d(u,v)$ is the shortest distance between u and v

$$\text{betweenness centrality } (w) = \frac{\sum\limits_{u<v} p_{uv}(w)/p_{uv}}{(n-1)(n-2)} \tag{3}$$

where $p_{uv}(w)$ is the number of shortest paths between u and v that pass through w

The objective is to conduct the social network analysis tasks from the information given in two social networks without violating the privacy, but still obtain the result as accurate as if it is conducted on the integrated social network. In this work, we propose the sub-graph generalization approach for information sharing and privacy protection and we focus on computing closeness centrality.

2 Related Work

The concerns on privacy of publishing data have been increasing in the recent years. Nowadays, many organizations are publishing data containing sensitive personal information for analysis. This data is usually represented as table, which includes medical, census, customer transactions, and voter registrations. This unaggregated information about individuals is important for trend analysis, medical research, and allocation of resources. However, individual privacy is always an important issue in the public. A solution to this issue is publishing distorted version of tables so that identifications cannot be easily detected.

A simple approach to such a version is removing attributes that uniquely identify a person such as names and identification numbers. However, trivial linking attack proves that this simple approach does not work. It is very likely to identify a person with his sensitive data by using the apparently innocuous sets of attributes such as age, gender, and zip code from other tables. These sets of attributes supporting linking attack are known as *quasi-identifiers*. An earlier study estimates that 87% of US population can be uniquely identified by linking attack [12]. According to Sweeney [13], it can identify the medical records of Massachusetts governor by linking several publishing tables. As illustrated, if one knows that Charles is a registered voter, he/she looks up Charles' registration record and uses the quasi-identifiers to cross-check with the hospital that he has visited. It is simple to identify that Charles has HIV which is his sensitive information (Table 2).

k-anonymity [11, 13] is the first approach to protect privacy in publishing data by generalizing the data set rather than only removing identification attributes. If every record in a table is indistinguishable from at least *k*–1 other records with respect to every set of quasi-identifier attributes, this table satisfies the property of *k*-anonymity. However, *k*-anonymity fails when there is a lack of diversity in the sensitive attributes or the one who attacks has other background knowledge. As illustrated, there is a lack of diversity on the attribute values of disease in the quasigroup with age = [6,8] and location = [00300,02000]. When one identifies Paul's attributes values of the quasi-identifiers in the voter registration records match with the values of the quasigroup in the medical records, it is easy to identify Paul's disease as viral infection.

Table 2 Medical records and voter registration records in information sharing using attribute removing and k-anonymity

Medical Records – Removing Names

Name	Age	Sex	Location	Disease
Peter	8	M	00330	Viral Infection
Paul	14	M	01540	Viral Infection
Andrew	18	M	18533	Viral Infection
Stephen	20	M	14666	Viral Infection
Charles	29	M	35667	HIV
Gordon	30	M	43986	Cancer
Linda	35	F	31147	Cancer
Mary	39	F	45246	Cancer
Stella	45	F	85103	Heart Disease
Angel	51	F	96214	HIV

quasi –identifiers

Voter Registration Records

Name	Age	Sex	Location
Charles	29	M	35667
Paul	14	M	01540
David	25	M	00338
…			

Medical Records – k-anonmity

Age	Sex	Location	Disease	
[5,20]	M	[00300,02000]	Viral Infection	quasi group A
	M		Viral Infection	
	M		Viral Infection	
	M		Viral Infection	
[20,40]	M	[20001,50000]	HIV	quasi group B
	M		Cancer	
	F		Cancer	
	F		Cancer	
[41,60]	F	[80000,99999]	Heart Disease	quasi group C
	F		HIV	

Voter Registration Records

Name	Age	Sex	Location
Peter	29	M	35667
Paul	14	M	00332
David	25	M	00338
…			

l-diversity [10] ensures that there are at least l well-represented values of the sensitive attribute for every set of quasi-identifier attributes. However, one can still estimate the probability of a particular sensitive value. As illustrated, quasigroup A is 1-diverse and therefore the attribute value of Disease of any records within this group can be identified as viral infection. However, quasigroups B and C are 2-diverse. As a result, we cannot identify the disease attribute value for any records within these two groups. Unfortunately, we can still estimate that the probability of the disease attribute values as cancer for any records in quasigroup B is 75%. If one has the background knowledge about Mary that she is HIV-negative in addition to matching the quasi-identifiers, it is easy for him or her to conclude that Mary's disease must be cancer.

Other enhanced techniques of k-anonymity and l-diversity have also been proposed recently. Personalized anonymity [18] allows a person to specify the degree of privacy protection for his sensitive values. Instead of publishing the exact sensitive value, it publishes a value in a higher level of the taxonomy of attribute that is acceptable to the user. (α,k)-Anonymity [17] allows a person to specify a threshold on the relative frequency of the sensitive data in every equivalence class.

The approaches of privacy protection on records presented in table formats are mainly focused on *domain generalization* [1]. Domain generalization partitions the values of a domain D into a number of partitions P_1, P_2, ... such that the union of these partitions is equal to the range of D. In general, the partitions do not have any overlapping, that means $\cup P_i = D$ and $P_i \cap P_j = \emptyset$. By partitioning the domains of several quasi-identifiers, it increases the difficulties for attackers to uniquely identify a person and his sensitive information.

The current research on privacy preservation of social network data (or graphs) focuses on the purpose of data publishing. A naïve approach to the preservation is removing the identities of all nodes but only revealing the edges of a social network. In this case, the global network properties are preserved for other research applications assuming that the identities of nodes are not of interest to the research applications. However, Backstrom et al. [2] proved that it is possible to discover whether edges between specific targeted pairs of nodes exist or not by active or passive attacks. Based on the uniqueness of small random sub-graphs embedded in a social network, one can infer the identities of nodes by solving a set of restricted isomorphism problems. Active attacks refer to planting well-structured sub-graphs in a published social network and then discovering the links between targeted nodes by identifying the planted structures. Passive attacks refer to identifying a node by its association with neighbors and then identifying other nodes that are linked to this association. Such attack can also be considered as neighborhood attack.

In order to tackle active and passive attacks and preserve the privacy of node identities in a social network, there are several anonymization models proposed in the recent literature: *k-candidate anonymity* [8],*k-degree anonymity* [9],*and k-anonymity* [24]. Such anonymization models are proposed to increase the difficulty of being attacked based on the notion of k-anonymity in tabular data. *k-candidate* anonymity [8] defines that there are at least k candidates in a graph G that satisfies a given query Q. k-degree anonymity [9] defines that, for every node v in a graph G, there are at least $k-1$ other nodes in G that have the same degree as v. k-anonymity [24] has the strictest constraint. It defines that, for every node v in a graph G, there are at least $k-1$ other nodes in G such that their anonymized neighborhoods are isomorphic. Zheleva [23] proposed an edge anonymization model for social networks with labeled edges rather than labeled nodes.

The technique to achieve the above anonymities is *edge or node perturbation* [8, 9, 24]. By adding and/or deleting edges and/or nodes, a perturbed graph is generated to satisfy the anonymity requirement. Adversaries can only have a confidence of $1/k$ to discover the identity of a node by neighborhood attacks.

Since the current research on privacy preservation of social network data focuses on preserving the node identities in data publishing, the anonymized social network can only be used to study the global network properties but may not be applicable to other SNAM tasks. In addition, the sets of nodes and edges in a perturbed social network are different from the sets of nodes and edges in the original social network. As reported by Zhou and Pei [24], the number of edges added can be as high as 6% of the original number of edges in a social network. A recent study [22] has investigated how edge and node perturbation can change certain network properties. Such distortion may cause significant errors in certain SNAM tasks such as centrality measurement although the global properties can be maintained.

3 A Framework of Information Sharing and Privacy Preservation for Integrating Social Networks

Figure 3 presents a framework of information sharing and privacy preservation for integration social networks. Assuming organization P (O_P) and organization Q (O_Q) have social networks G_P and G_Q respectively, O_P needs to conduct a SNAM task, but G_P is only a partial social network for the SNAM task. If there is not any privacy

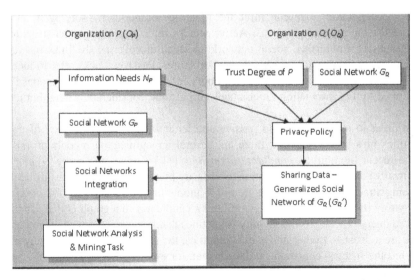

Fig. 3 A proposed framework of information sharing and privacy preservation for integrating social networks

concern, one can integrate G_P and G_Q to generate an integrated G and obtain a better SNAM result. Due to privacy concern, O_Q cannot release G_Q to O_P but only shares some data with O_P according to the agreed privacy policy. At the same time, O_P does not need all data from O_Q but only those that are critical for the SNAM task. The objectives are maximizing the information sharing that is needed for the SNAM task but conforming to the privacy policy so that sensitive information can be preserved and more accurate SNAM result can be obtained.

The information we share can be sensitive or insensitive and useful or not useful for a SNAM task to be conducted by the information requesting party. In this proposed project, we need to maximize the information sharing that is insensitive and useful for the SNAM tasks. The shared information should not include any sensitive information; however, it must be useful for improving the performance of the SNAM task conducted by the information requesting party.

4 Sub-graph Generalization

A social network is not represented as a table, but can be represented as an adjacency matrix. Domain generalization is not applicable because we don't have a domain of any particular attribute. The current anonymization and perturbation approach to social network privacy preservation can only support the study of global network properties of the published social network but does not support the integration of social networks for analysis. We propose an approach in privacy protection of social networks by generalizing sub-graphs of a social network. By generalizing a sub-graph, we want to protect the identification of nodes that are not known in public, the structure of the sub-graph, and the types of relationships among the nodes within the sub-graph. However, generalizing a sub-graph will lose the important information for social network analysis when a social network is integrated with other social networks.

In order to maintain a certain degree of utility of a generalized sub-graph, we provide other general information about sub-graphs for the purpose of analysis. Such general information should not reduce the degree of privacy; on the other hand, it should provide additional utility for social network analysis. These general information are the general properties of graphs such as number of nodes in a sub-graph, length of shortest paths, and possible types of relationships.

Given a social network, $G = \{V, E\}$, the sub-graph generalization approach partitions G into several connected sub-graphs $G_1, G_2, \ldots$ Each sub-graph has a set of nodes and a set of edges, $G_i = \{V_i, E_i\}$. There are edges that connect G_i together as the original graph G. E_{ij} denotes the edge connecting G_i and G_j. Since these partitioned sub-graphs do not have any overlapping (i.e., $V_i \cap V_j$), the union of the nodes from all the sub-graphs is the original set of nodes (i.e., $\cup V_i = V$). The union of the edges from all the sub-graphs and other connecting edges of the sub-graphs is the original set of edges (i.e., $\cup E_i + \cup E_{ij} = E$).

Suppose there are n insensitive nodes in graph G, we outline some potential methods that divide G into n sub-graphs. Each sub-graph contains a unique insensitive node as the center of the sub-graph.

1. Treat each insensitive node as center of a sub-graph. Each sensitive node will be assigned to the nearest insensitive node to form a sub-graph.
2. Treat each insensitive node as center of a sub-graph. Each sensitive node will be assigned to one particular sub-graph in order to minimize the sum of radius of all sub-graphs.
3. Treat each insensitive node as center of a sub-graph. Each sensitive node will be assigned to one particular sub-graph in order to minimize the sum of diameter of all sub-graphs.
4. Our work can also be extended to weighted graph. Treat each insensitive node as center of a sub-graph. Each sensitive node will be assigned to one particular sub-graph in order to minimize the sum of weight of minimum spanning tree in each sub-graph.

The partitioned sub-graph (or known as generalized node) is represented as a node in the generalized graph. The total number of generalized nodes in the generalized graph equals the number of partitioned sub-graphs. The total number of edges in the generalized graph equals the number of connecting edges of sub-graphs, $G = \{\{\ldots, G_i, \ldots\}, \{\ldots, E_{ij}, \ldots\}\}$.

The adjacency matrix of the generalized graph is transformed by merging the cells that represent the relationships of the nodes within a generalized sub-graph or between generalized sub-graphs. The adjacency matrix of a generalized graph with K generalized sub-graphs has $K \times K$ merged cells. The value in a cell between two generalized sub-graphs represents the type of relationship between the representing nodes of the sub-graphs (i.e., the node of the known identification in public). The values in the cell between the same sub-graphs represent the generalized information of the generalized sub-graph. E_{ij} exists between G_i and G_j if there is at least one edge between the nodes of G_i and the nodes of G_j, otherwise, E_{ij} does not exist.

5 Policy of Sharing

In our framework as presented in Fig. 3, we propose the development of privacy policy by considering the information needs based on the SNAM task, the trust degree of the information requesting party, and the information available in its own social network. The privacy policy determines what data can be shared. Thuraisingham [15, 16] discussed a coalition of dynamic relational data sharing, in which security and integrity policies are enforced. When we perform social network data sharing, we need to consider what kinds of nodes and edges are needed to accomplish a particular SNAM task by analyzing the network structure. The trust degree of the information requesting party will determine certain sensitive data to be protected and certain insensitive data to be shared. At the same time, the information available for sharing may not be able to obtain the exact SNAM result, but our objective is maximizing the accuracy so that a better result can be obtained in comparing with the SNAM result obtained from the original network without sharing.

As reported in the Washington Post in September 2008, there was no systematic mechanism for sharing intelligence between private companies or between companies and the government. It also emphasized that the government should take actions on developing mechanism to share unclassified information while some information should remain classified. Without information sharing, American-developed products and technology could be easily stolen with little effort. The key point is differentiating the sensitive and insensitive information to permit necessary information sharing while protecting the privacy. A well-developed privacy policy provides a mechanism to determine what information should be shared based on the information needs and the trust degree of the information requesting party.

When a policy is developed to determine the information to be shared in a generalized node, there are a number of criteria to be considered [16]. We give a few examples that are considered to be important.

> *Need to know*: Some information is essential for conducting a social network analysis but some other information is not as important depending on the types of analysis. As a result, the information that needs to be shared depends on what the information requester needs to do. For example, if the information requester is computing closeness centrality of a criminal social network, information such as $P(w)$ and R are not necessary.
>
> *Need to share*: Some information is crucial for discovering knowledge. Without such, there is very limited knowledge one can discover.
>
> *Privacy*: The private information of individuals that is under a high risk of violating laws or endangering the person involved should not be disclosed.
>
> *Trust*: Information providers may have different degrees of trust on different information requesters. The amount of information to be shared can be adjusted by the degree of trust. The complete information can be disclosed to an information requester who has a hundred percent trust from the information provider, regardless of the need. On the other hand, the information provider may decide not to disclose any information if the information requester has zero percent trust from itself.

6 Shared Information in a Generalized Node

Information available for sharing in a generalized node in a generalized graph determines the utility of the generalization. In order to conform to the policy of sharing, identities of non-public persons and their relationships cannot be released. However, general information such as the number of nodes in a generalized node and the lengths of shortest paths do not violate the privacy information of individuals.

We propose publishing a combination of the following information in a generalized node depending on the intended degree of privacy.

1. Number of nodes in a social network or a subgroup of a social network
2. The connecting nodes between two subgroups of a social network
3. The maximum/average/minimum length of the shortest path in a social network or a subgroup of a social network
4. The maximum/average/minimum degree of the nodes in a social network or a subgroup of a social network
5. The number of shortest paths going through an insensitive node between any two nodes in a social network or a subgroup of a social network
6. The length of the shortest path between two insensitive nodes
7. The number of the shortest paths between two insensitive nodes
8. The degree of the insensitive nodes
9. The eccentricity/k-centricity of the insensitive nodes in a social network or a sub-graph of a social graph
10. The radius/diameter of a social network or a sub-graph of a social graph
11. The center/peripheral/median of a social network or a sub-graph of a social graph if it is an (or they are) insensitive node(s)
12. The eccentricity of a sub-graph in a social network

7 Computation with Shared Information

If we don't know the connecting nodes between two generalized nodes, we can compute the maximum, minimum, and average distances between any pairs of nodes in any two generalized nodes as follows:

$$d^{\max}(u,v) = \begin{cases} L\_SP(U) + L\_SP(V)+ \\ \sum_{\forall W} L\_SP(W) + |W| + 1 & \text{if } U \neq V \\ L\_SP(U) & \text{else} \end{cases} \tag{4}$$

$$d^{\min}(u,v) = \begin{cases} |W| + 1 & \text{if } U \neq V \\ 1 & \text{else} \end{cases} \tag{5}$$

$$d^{\text{avg}}(u,v) = \begin{cases} AVG\_SP(U) + AVG\_SP(V)+ \\ \sum_{\forall W} AVG\_SP(W) + |W| + 1 & \text{if } U \neq V \\ AVG\_SP(U) & \text{else} \end{cases} \tag{6}$$

where

> U and V are sub-graphs (generalized nodes) in a generalized graph,
> $u \in U, v \in V$
> W is a sub-graph (generalized node) in the shortest path between u and v in a generalized graph
> $L\_SP(W)$ is the maximum length of the shortest paths between any two nodes in a generalized node, W
> $AVG\_SP(W)$ is the average length of the shortest paths between any two nodes in a generalized node, W.

If we know the connecting nodes of a sub-graph W which is in the shortest path between u and v in a generalized graph and we know the length of the shortest path between these two nodes SP, we can replace the maximum length of the shortest path between any pairs of nodes in the sub-graph $L\_SP(W)$ and the average length by the shortest path between the two known connecting nodes.

$$L\_SP(W) \text{ is substituted by } SP(a,b)$$

where

> a is a known connecting node between W and the sub-graph before W in the shortest path between u and v in a generalized graph and
> b is a known connecting node between W and the sub-graph after W in the shortest path between u and v in a generalized graph.

If the known connecting nodes of a sub-graph which is in the shortest path between u and v in a generalized graph and these connecting nodes in a sub-graph are indeed the same node, then $L\_SP(W)$ can be replaced by 0.

In the special case that the connecting nodes are the same in all sub-graphs along the shortest path between u and v, $d^{\max}(u,v)$ equals to $d^{\min}(u,v)$.

Similarly, we replace $AVG\_SP(W)$ by $SP(a,b)$ or 0 in computing $d^{\mathrm{avg}}(u,v)$ when the shortest path between the known connecting nodes in W is available or the connecting nodes in W are the same.

As we have discussed before, we can share more than maximum, average and minimum length of the shortest path for one sub-graph. If we generate a sub-graph by treating the insensitive node as the center, we can share information such as radius $R(\cdot)$ and eccentricity $E(\cdot)$ of the center of each sub-graph. We define the distance, d^{mix}, between any pairs of nodes in any two generalized nodes as follows:

(1) From one insensitive node u to another insensitive node v

$$d^{\mathrm{mix}}(u,v) = R(u) + R(v) + \sum_{\forall W} L\_SP(W) + |W| + 1 \qquad (7)$$

(2) From one insensitive node u to one sensitive node v

$$d^{\text{mix}}(u,v) = \begin{cases} R(u) + L\_SP(V) + \sum_{\forall W} L\_SP(W) + |W| + 1 & \text{if } U \neq V \\ E(u) & \\ & \text{else} \end{cases} \quad (8)$$

(3) From one sensitive node u to another sensitive node v

$$d^{\text{mix}}(u,v) = \begin{cases} L\_SP(U) + L\_SP(V) + \\ \sum_{\forall W} L\_SP(W) + |W| + 1 & \text{if } U \neq V \\ L\_SP(U) & \text{else} \end{cases} \quad (9)$$

where

U and V are sub-graphs (generalized nodes) in a generalized graph,
$u \in U, v \in V$
$R(u)$ is the radius of sub-graph U
W is a sub-graph (generalized node) in the shortest path between u and v in a generalized graph
$L\_SP(W)$ is the maximum length of the shortest paths between any two nodes in a generalized node, W.

Example 2: G_2 in Example 1 can be generalized to a graph with four generalized nodes based on what we want to further analyze from what we know about the structure of G_1. Table 3 shows the adjacency matrix of G_2.

Table 3 Adjacency matrix of G_2

	A	B	C	D	E	F	G	H
A	0	1	0	0	0	0	0	0
B	1	0	1	1	0	1	0	0
C	0	1	0	0	0	1	0	0
D	0	1	0	0	1	0	0	0
E	0	0	0	1	0	0	0	0
F	0	1	1	0	0	0	1	1
G	0	0	0	0	0	1	0	0
H	0	0	0	0	0	1	0	0

Table 4 presents the adjacency matrix of the generalized graph of G_2 with four generalized nodes. The first node generalizes the sub-graph of two nodes, A and B. The second node is C. The third node generalizes the sub-graph of D and E. The fourth node generalizes F, G, and H.

Table 4 Adjacency matrix of the generalized graph of G_2

	A	B	C	D	E	F	G	H
A	$N=2$		1	1		1		
B	$L\_SP=1$							
C	1		$N=1$ $L\_SP=0$	0		1		
D	1		0	$N=2$		0		
E				$L\_SP=1$				
F	1		1	0		$N=3$		
G						$L\_SP=2$		
H								

Fig. 4 The generalized graph of G_2

Table 5 Distance between A and other nodes and closeness centrality of A as computed from G_1, G, and the information shared from the generalized graph of G_2

	G_1	G	\multicolumn{4}{Information sharing from the generalized graph of G_2}			
			d^{max}	d^{avg}	d^{min}	d^{mix}
$d(A,B)$	1	1	1	1	1	1
$d(A,C)$	2	2	2	2	2	2
$d(A,D)$	3	2	3	3	2	3
$d(A,E)$	4	3	3	3	2	3
$d(A,F)$	4	2	4	3.33	2	3
$d(A,G)$	5	3	4	3.33	2	4
$d(A,H)$	6	3	4	3.33	2	4
Closeness	7/25	7/16	7/21	7/19	7/13	7/20
centrality(A)	=	=	=	=	=	=
	0.28	0.44	0.33	0.37	0.54	0.35

The generalized graph of G_2 is presented in Fig. 4. The generalized node is labeled by one of the known node in the sub-graph.

The second and third columns of Table 5 show the distance computed from G_1 and G. G is the integrated graph of G_1 and G_2. The closeness centrality of A computed from G_1 has an error of 42.9% taking the closeness centrality of A computed from G. However, if we use the shared information from the generalized graph of G_2, the error can be reduced to 11% using d^{avg} for computation.

8 Experiment

We have conducted an experiment to evaluate the effectiveness of using the sub-graph generalization approach for information sharing between social networks to compute the closeness centrality. In this experiment, we randomly generated ten pairs of graphs. For each pair of graphs, we created an integrated graph. We computed the closeness centralities of the nodes in the integrated graph as the benchmark. By using the first graph in each pair of graphs, we computed the closeness centralities of the nodes in this graph and computed the errors. We then generalized the second graph and used the shared information to compute the closeness centralities of the same set of nodes and the errors again. For the ten pairs of graphs, the average error of closeness centrality computed from the first graph without information sharing is 35%. The average error of closeness centrality computed with shared information of the second using $d^{\max}$, d^{avg}, and $d^{\min}$ is 23, 17, and 28%, respectively. It shows that the sub-graph generalization approach of information sharing between social networks improves the social network analysis such as closeness centrality computation substantially. We have not tested d^{mix} yet but we shall further investigate the usage of insensitive nodes in social network integration in our future work.

9 Conclusion

We have proposed a sub-graph generalization approach for sharing information between terrorist or criminal social networks. The sub-graph generalization tries to retain the degree of privacy in information sharing and yet shares information with high utility so that an effective social network analysis can still be conducted. In our experiment, it shows that the proposed technique is promising to reduce the error substantially when we compare the computation of closeness centrality without and with information sharing by sub-graph generalizations. In the future, we shall further investigate other generalization techniques and determine what information to be shared under different circumstances or analysis.

References

1. Aggarwal G, Feder T, Kenthapadi K, Khuller S, Panigrahy R, Thomas D, and Zhu A (2006) Achieving Anonymity via Clustering. In: Proceedings of PODS. June 26–28, Chicago, IL.
2. Backstrom L, Dwork C, and Kleinberg J (2007) Wherefore Art Thou R3579X? Anonymized Social Networks, Hidden Patterns, and Structural Steganography. In: Proceedings of the 16th International WWW'07 Banff, Alberta.
3. Baird Z, Barksdale J, and Vatis M (2003) Creating a Trusted Network for Homeland Security. Markle Foundation.
4. Baird Z and Barksdale J (2006) Mobilizing Information to Prevent Terrorism: Accelerating Development of a Trusted Information Sharing Environment. Markle Foundation.

5. Best C, Piskorski J, Pouliquen B, Steinberger R, and Tanev H (2008) Automating Event Extraction for the Security Domain. Intelligence and Security Informatics-Techniques and Applications. Editors: Chen H, Yang CC, Berlin (Germany): Springer Verlag, p. 17–43.
6. Caruson K, Macmanus SA, Khoen M, and Watson TA (2005) Homeland Security Preparedness: The Rebirth of Regionalism. Publius 35(1): 143–189.
7. Friedmann RR and Cannon WJ (2005) Homeland Security and Community Policing: Competing or Complementing Public Safety Policies. Journal of Homeland Security and Emergency Management 4(4): 1–20.
8. Hay M, Miklau G, Jensen D, Weis P, and Srivastava S (2007) Anonymizing Social Networks. Technical Report 07-19. University of Massachusetts, Amherst.
9. Liu K and Terzi E (2008) Towards Identity Anonymization on Graphs. In: ACM SIGMOD'08. Vancouver, BC, Canada: ACM Press.
10. Machanavajjhala A, Gehrke J, Kifer D, and Venkitasubramaniam M (2006) l-Diversity: Privacy Beyond k-Anonymity. In: Proceedings of the 22nd International Conference on Data Engineering, Atlanta, GA.
11. Samarati P (2001) Protecting Respondents' Identities in Microdata Release. IEEE Transactions on Knowledge and Data Engineering 13(6): 1010–1027.
12. Sweeney L (2000) Uniqueness of Simple Demographics in the US Population. Technical Report. Carnegie Mellon University.
13. Sweeney L (2002) k-Anonymity: A Model for Protecting Privacy. International Journal on Uncertainty Fuzziness Knowledge-Based Systems 10(5): 557–570.
14. Thacher D (2005) The Local Role in Homeland Security. Law & Society 39(3): 557–570.
15. Thuraisingham B (1994) Security Issues for Federated Databases Systems. Computers and Security, North Holland, December.
16. Thuraisingham, B (2008) Assured Information Sharing: Technologies, Challenges and Directions. Intelligence and Security Informatics-Techniques and Applications. Editors: Chen H, Yang CC, Berlin (Germany): Springer Verlag, p. 1–15.
17. Wong RC, Li J, Fu A, and Wang K (2006) (α,k)-Anonymity: An Enhanced k-Anonymity Model for Privacy-Preserving Data Publishing. In: Proceedings of SIGKDD. August 20–23, Philadelphia, PA.
18. Xiao X and Tao Y (2006) Personalized Privacy Preservation. In: Proceedings of SIGMOD. June 27–29, Chicago, IL.
19. Yang CC, Liu N, and Sageman M (2006) Analyzing the Terrorist Social Networks with Visualization Tools. In: Proceedings of the IEEE International Conference on Intelligence and Security Informatics. May 23–24, San Diego, CA.
20. Yang CC and Sageman M (2009) Analysis of Terrorist Social Networks with Fractal Views. Journal of Information Science 35(3): 299–320.
21. Yang CC and Ng TD (2007) Terrorism and Crime Related Weblog Social Network: Link, Content Analysis and Information Visualization. In: Proceedings of the IEEE International Conference on Intelligence and Security Informatics. May 23–24, New Brunswick, NJ.
22. Ying X and Wu X (2008) Randomizing Social Networks: A Spectrum Preserving Approach. In: SIAM International Conference on Data Mining (SDM'08). Atlanta, GA.
23. Zheleva E and Getoor L (2007) Preserving the Privacy of Sensitive Relationships in Graph Data. In: First ACM SIGKDD International Workshop on Privacy. Security, and Trust in KDD (PinKDD'07), San Jose, CA.
24. Zhou B and Pei J (2008) Preserving Privacy in Social Networks against Neighborhood Attacks. In: IEEE International Conference on Data Engineering, Atlanta, GA.

Processing Constrained k-Closest Pairs Queries in Crime Databases

Shaojie Qiao, Changjie Tang, Huidong Jin, Shucheng Dai, Xingshu Chen, Michael Chau, and Jian Hu

Abstract Recently, spatial analysis in crime databases has attracted increased attention. In order to cope with the problem of discovering the closest pairs of objects within a constrained spatial region, as required in crime investigation applications, we propose a query processing algorithm called Growing Window based Constrained k-Closest Pairs (GWCCP). The algorithm incrementally extends the query window without searching the whole workspace for multiple types of spatial objects. We use an optimized R-tree to store the index entities and employ a density-based range estimation approach to approximate the query range. We introduce a distance threshold with regard to the closest pair of objects to prune tree nodes in order to improve query performance. Experiments discuss the effect of three important factors, i.e., the portion of overlapping between the workspaces of two data sets, the value of k, and the buffer size. The results show that GWCCP outperforms the heap-based approach as a baseline in a number of aspects. In addition, GWCCP performs better within the same data set in terms of time and space efficiency.

Shaojie Qiao

School of Information Science and Technology, Southwest Jiaotong University, Chengdu 610031; Southwest Jiaotong University

e-mail: qiaoshaojie@gmail.com

Changjie Tang, Shucheng Dai, Xingshu Chen, Jian Hu*

School of Computer Science, Sichuan University, Chengdu 610065, China

e-mail: {cjtang,daishucheng,chenxsh}@scu.edu.cn;
*hujianlucky@163.com

Huidong Jin

Mathematical and Information Sciences, CSIRO, ACT 2601, Australia

e-mail: warren.jin@csiro.au

Michael Chau

School of Business, The University of Hong Kong, Pokfulam, Hong Kong, SAR

e-mail: mchau@business.hku.hk

An earlier version of the paper received the best paper award at the IEEE International Conference on Intelligence and Security Informatics 2008.

Keywords Spatial analysis · Crime databases · Constrained closest pairs · Query processing · R-tree

1 Introduction

Crime databases have developed into an important tool in crime investigate applications. In general, a "crime database" is defined as a spatial database that stores incident-based data focusing on the unique characteristics of a criminal incident that captures detailed crime characteristics, e.g., location, modus operandi, and time [1]. The locations of crime incidents illustrated in a crime-based map are represented by 2D points or 3D geometric entities.

Query processing for crime data has recently become popular, since the tragic events of September 11 and the subsequent anthrax contamination of letters produced a great effect on many aspects of society [2]. Crime investigators and anti-terrorism specialists typically maintain a spatial database storing the locations, distances, time, and other relevant information of crimes. For example, given the description of a fat suspect with long beard, a round face, and brown hair, an investigator is apt to find the closest pairs of spatial objects such as subways and airports where this criminal frequently visits. It is of practical value for law enforcement agencies to develop an efficient spatial query processing method in crime databases for predicting the potential crimes and preventing future crimes.

In addition, crime authorities can employ crime databases to allocate new facilities more appropriately. For instance, emergency calls to an emergency call center are dialed from various locations. For each call, a staff can generate a spatial event associated with an accident and dispatch ambulances to the scene of accident that is nearest to them. Another example is that the police station of New York City used its GIS to locate facilities and to respond to emergencies such as the attack on the World Trade Centers, in which location-based service plays an essential role in response and recovery efforts [3].

As for security domain, it is important to discover the closest pairs of multiple types of objects (i.e., objects derived from distinct spatial data sets) in crime databases. For example, a policeman may need to find one or multiple closest pairs of roads and crime scenarios, instead of the closest pairs of roads, in an efficient and effective manner.

A very common spatial query is the "k-Nearest Neighbor Query" (kNN) [4]. For example, a staff may want to find the k ($k \geq 1$, is the number of nearest neighbors) police officers closest to a crime scenario where a crime with k injured people occurred. The problem of "k-Closest Pairs Query" (k-CPQ) is an extension by combining nearest neighbor query with spatial join in order to find the k-closest pairs of spatial objects from two distinct data sets [5]. For example, anti-terrorist officers may be interested in finding the closest five supermarkets and banks, or the closest four bus stops and railways.

In this study, the *k*-CPQ with a spatial constraint is called constrained *k*-closest pairs query (*k*-CCPQ) [6]. This problem is of great practical value when applied to the security informatics domain of crime databases due to the following reasons.

- Typically, police officers often care for the query results within a given area which can help save query time instead of the whole space in the real-life scenarios.
- *k*-CCPQs may ask for *k*-closest pairs of any two types of objects. It is more appropriate for crime databases, since crime databases consist of multiple types of objects. For instance, the *k*-CCPQs in crime databases discovering the *k* pairs of subways and police departments resort to find the *k*shortest distances so that the police officer can efficiently dispatch police officers to the subways which are the possible targets of terrorist attack. In other words, it can assist in making an optimal allocation of police resources.

In order to solve the *k*-CCPQ problem in crime databases, we make the following contributions in this study:

1. We propose a novel constrained *k*-closest pairs query processing algorithm based on growing windows in crime databases, namely GWCCP. The window extends incrementally and terminates when discovering the *k*-CCPQs, which help eliminate the unnecessary distance calculations between spatial objects of crime incidents or other point locations.
2. We employ an optimized R-tree index structure to store the index entities and treat the distance of the maintained closest pair as a threshold. The closest pair whose distance is greater than this threshold is pruned which benefits reducing the response time of tracing criminals.
3. We use a density-based range estimation method to compute the square query range. It has some advantages. First, the space required to store the density information takes only several bytes. Second, every time a new range estimate is required, it is derived from the density of the previous window for query.
4. We conduct experiments to compare the proposed approach with the heap-based algorithm for *k*-CCPQ over two distinct crime data sets, and the results show that our algorithm performs well in most cases. In particular, the performance of GWCCP is better than SRCP-tree [7] within the same data set.

The rest of this chapter is organized as follows. Section 2 surveys the related work. Section 3 describes the problem of *k*-CCPQ and presents some useful metrics. Section 4 introduces the density-based range estimation method and proposes the algorithm of *k*-CCPQ query. Section 5 presents the performance studies of the proposed algorithm and discusses the experimental results. Finally, Section 6 concludes this chapter with a summary and directions for future work.

2 Related Work

The closest pairs query problem has unleashed a new wave of applications in the research of spatial databases [4–6]. However, there is relatively little work that is relevant to the closest pairs query processing problem in crime databases that is important to the security informatics domain.

Much work on the closest pair query problem focuses on applying R-tree to kNN queries, because R-tree is an efficient spatial index structure of retrieving data items based on the location of each object [8].

Roussopoulos [9] proposed an R-tree algorithm for kNN queries. The disadvantage of the algorithm is that once a node is visited, all nodes in its sub-tree have to be visited as well.

To avoid direct accesses to spatial indices, Liu et al. [10] transformed a kNN query into more window queries.

Hjaltason and Samet [11] proposed two spatial join operations between two R-tree indices. However, the approach is still required to store every pair of index entries and spatial objects in a priority queue. Thus, it still cannot avoid a large number of disk accesses.

Corral proposed an improved approach known as the Heap algorithm [4]. But, this approach is not efficient when there is much overlap between two spatial data sets, and this study does not take into account the case of k-CPQ with spatial constraints.

Ferhatosmanoglu et al. [12] applied some methods to answer the constrained nearest neighbor queries. However, the proposed algorithms are not suitable for k-CCPQ query.

The most similar work to this study has been explored by Shan. He made a good attempt to solve the k-CCPQ problem and proposed two kinds of SRCP-tree (Self Rang Closest Pair tree) [7]. However, SRCP-tree cannot support the CP (Closest Pair) query for multiple types of spatial objects.

To cope with the k-CCPQ problem in crime databases, we will introduce an efficient k-CCPQ query processing algorithm in the following section. This method can be applied to other spatial databases as well.

3 Problem Description

In this study, the k-CCPQ problem in crime databases is to seek k pairs of crime sites in two distinct data sets, and the sites located within a given spatial constraint. The formal definition is shown below [4].

Definition 1 (k-CCPQ) Given two spatial data sets, $S = \{s_1, s_2, \ldots, s_M\}$ and $T = \{t_1, t_2, \ldots, t_N\}$, be stored in two R-trees T_M and T_N, respectively. The k-CCPQ of S and T with regard to a given spatial constraint R is defined as k-ordered pairs as follows:

$$(s_{l_1},t_{h_1}),(s_{l_2},t_{h_2}),\ldots,(s_{l_k},t_{h_k})$$

where $s_{l_1},s_{l_2},\ldots,s_{l_k} \in S, t_{h_1},t_{h_2},\ldots,t_{h_k} \in T$, and each $s_i \in S$ and $t_i \in T$ has similar characteristics, respectively, $(s_{l_1},s_{l_2},\ldots,s_{l_k})$ and $(t_{h_1},t_{h_2},\ldots,t_{h_k})$ are inside R, such that:

$$\text{dist}\left(s_i,t_j\right) \geq \text{dist}\left(s_{l_k},t_{h_k}\right) \geq \text{dist}\left(s_{l_{k-1}},t_{h_{k-1}}\right) \geq \ldots \geq \text{dist}\left(s_{l_1},t_{h_1}\right)$$

$$\forall \left(s_i,t_j\right) \in (S \times T\text{-}\{(s_{l_1},t_{h_1}), (s_{l_2},t_{h_2}),\ldots, (s_{l_k},t_{h_k})\}).$$

The k-CCPQs from the Cartesian product of S and T within *Rare* k pairs that have the shortest distances between all pairs of points that are formed by selecting one point from S and the other point from T. Although "dist" stands for Euclidean distance in this study, the proposed method can be easily adapted to Minkowski distance as well. Here, we will give the useful metrics to measure the distances between two spatial objects.

Given two MBRs for S and T, two spatial objects $s_i \in S$ and $t_j \in T$. Following [4], MinMinDist(S, T) is the shortest distance between S and T boundaries and defined as:

$$\text{MinMinDist}(S,T) = \min\{\text{MiDist}(s_i,t_i)\} \tag{1}$$

where MinDist(s_i, t_i) represents the shortest distance between s_i and $\bar{t}_i$.

MaxMaxDist(S, T) is the maximum distance between two points falling on S and T boundaries. MinMaxDist(S, T) is the minimum distance that guarantees that there is at least one pair of objects with distance smaller or equal to MinMaxDist [4]. They are defined below [4].

$$\text{MaxMaxDist}(S,T) = \max\{\text{MaxDist}(s_i,t_i)\} \tag{2}$$

$$\text{MinMaxDist}(S,T) = \min\{\text{MaxDist}(s_i,t_i)\} \tag{3}$$

where MaxDist(s_i, t_i) represents the maximum distance between s_i and t_i.

In crime databases, the MBRs of two data sets frequently overlap, because the criminals often commit an offense in a similar location. It is evident that the higher the portion of overlapping between two MBRs, the higher the probability that more pairs with small distances appear [4]. So, we propose a k-CCPQ processing algorithm especially suitable for handling the case of higher overlap between two data sets in this study.

4 *k*-CCPQ Processing Based on Growing Windows

This section introduces a new algorithm for *k*-CCPQ in crime databases called
GWCCP by combining a new R-tree derived from SRCP-tree [7] to store the index
entities and a density-based range estimation method [10]. The tree structure used
in this paper is called C-tree. The main differences from the existing R-tree based
index structures, such as Guttman's linear and quadratic R-tree [8], are on the following aspects [6]: (a) each index entry i is augmented with a triple $(r_1, r_2, dist)$,
where r_1 and r_2 are the closest pair of objects in the sub-tree rooted by i and $dist$ is the
distance between r_1 and r_2; (b) C-tree uses the Least Recently Used (LRU) buffer
policy [13] in spatial selection, and spatial join between distinct data sets.

   The spatial constraint is represented by $win = [(x_l, y_l); (x_u, y_u)]$, where (x_l, y_l)
and (x_u, y_u) are the lower-left and upper-right corners of the spatial constraint. In
this study, each object is represented by a 2D point.

Definition 2 (*k*-CCPQ on Windows) Given a set of spatial objects denoted as P,
and a window w in a spatial constraint R, the *k*-CCPQ on windows refers to find the
k-closest pairs of objects from P located in w.

   For instance, consider the query: "find the four pairwise closest airports and
hotels located in a specified city." For this problem, we set a street zone in this
city as a query window, and expand it to find the required CPs.

   Figure 1 gives an illustrative example of the *k*-CCPQ problem on windows, where
the first data set is represented by stars while the second data set by crosses. Here,
the window R depicted by the dashed line is a spatial constraint and the window
w represented by the dashdotted line is an initial query window. In Fig. 1, we can
observe that the 1-RCPQ is (p_2, q_1), and the 2-RCPQs are (p_2, q_1) and (p_2, q_2).
Similarly, it is easy to find the 3-RCPQs, 4-RCPQs, etc.

Fig. 1 Example of the
k-CCPQs on window query
problem

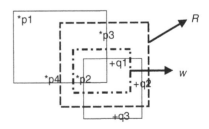

   Some useful notations are given here. Let S and T be two spatial data sets, s be
an index entry pointing to some node in a C-tree, $Node(s)$ be the node that s points
to, and $Sub\_tree(s)$ is the sub-tree rooted by $Node(s)$. The C-tree stores the closest
pair information by a triple $(o_1, o_2, dist)$ along with the index entry pair (p_1, p_2).
(o_1, o_2) is the closest pair of objects, where o_1 and o_2 come from objects indexed

by $Sub\_tree(p_1)$ and $Sub\_tree(p_2)$, respectively. In a C-tree, we borrow the buffer model proposed by Bhide [13] to manage the node update operations.

4.1 Node Insertion

C-tree uses the regular R-tree insertion algorithm to insert objects [7]. Let $\{(s_1, t_1), (s_2, t_2), \ldots, (s_n, t_n)\}$ be the index entry pairs pointing to the tree nodes that are along the insertion path, where s_1 points to the root node of S, s_n points to the leaf node that represents a newly inserted object, and $Node(s_i)$ is a parent node of $Node(s_{i+1})$.

(s_i, t_i) needs to be updated iff there is an object r in the sub-tree rooted by $Node(s_i)$, and an object $r\bullet$ in the sub-tree rooted by $Node(t_i)$ satisfying dist(r, $r\bullet$) < dist(s_i, t_i). If it finds such an object $r\bullet$, then update the closest pair; otherwise, (s_i, t_i) keeps unchanged.

4.2 Node Update

If an object is changed, its index record must be deleted, updated, and then re-inserted. The node update algorithm is shown as follows:

Algorithm 1: Node Update

Input: two C-tree A, B
Output: two updated C-tree A', B'
Method:

1. **For** (every object p in A)
2.   **If** (there is a path from the root node to p)
3.   Update the node along the path;
4.   **End if**
5.   **If** (the object in A does not have the node in the same level)
6.   Update the leaf node in B;
7.   **End if**
8. **End for**
9. **For** (every object q in B)
10.   Apply the similar update operation as the object in A;
11. **End for**

In step 2, we update the node as finding the node in the same level in B by using the closest pair computation method of SRCP-tree [7]. Note, for the node deletion operation, if p is a leaf node, C-tree uses the plane-sweep algorithm [14] to find the new closest pair of objects.

4.3 Query Processing

In this section, we propose a new window query algorithm, namely WinQuery [6], to find the k-CCPQs, and borrow the idea of density-based kNN query algorithm [10] to this approach.

First, we use the EstiRange1 function [10] to approximate the query radius. The difference from EstiRange1 function is that GWCCP uses the square query method instead of the circle query used in [11]. The initial radius r_0 can be calculated by the following equation:

$$r_0 = \sqrt{\frac{k(x_u - x_l)(y_u - y_l)}{N}} \qquad (4)$$

The query range W in the intersecting portion between the spatial constraint R and two existing spatial data sets S and T needs to be calculated first. The x-axis value of the upper-right corner of W is computed by Equation 5. Similarly, this equation can be applied to compute x_l, y_l, and y_u.

$$W.x_u = \min\{R.x_u, \max\{S.x_u, T.x_u\}\} \qquad (5)$$

Second, we obtain the results from the estimated range by calling the WinQuery algorithm as shown in Algorithm 2 and the results are inserted into a temporary queue *temp*. θ, is a threshold that is used to determine whether to add a node or not, and its initial value is set as $\theta = \infty$, which represents a sufficiently large integer.

Then, create an empty priority query *priority* to store the closest pairs and find the k-CPQs within the new window. The algorithm terminates when *count* (the number of closest pairs found so far) is larger than or equal to k; otherwise, closest pairs have to be further obtained by gradually extending the window. The growing range is computed from the current window by the EstiRange2 function [10]. EstiRange2 returns a query radius denoted as r_n. The function used in this study is distinct from the one proposed in [10]. When *count* = 0, we use the factor of 1.5 (that is an empirical value by experiments) instead of 2 to expand the radius. When *count* $\in$ (0, k–1], the denominator under the radical sign does not have the factor of π, this is because we use square query. The query radius is defined in Equation 6.

Algorithm 2: WinQuery(C-tree S, C-tree T, Window w)

1. Create an empty queue *temp*;
2. **If** (both objects from S and T in the closest pair are inside w)
3.   Put the closest pair into a priority queue *queue*;
4. **End if**
5. **While** (*queue* is not empty)
6.   Pops one triple (e_1, e_2, *dist*) from *queue*;
7.   **If** (*dist* > θ)
8.   Continue;

9.   **End if**
10. **If** (both $Node(e_1)$ and $Node(e_2)$ are leaf nodes)
11.   **For** (every object $se_1 \in Node(e_1)$ and $se_2 \in Node(e_2)$ in w)
12.   **If** (the distance between these two objects is smaller than θ)
13.   Update *temp* and $\theta = $ MinMinDist(se_1, se_2);
14.   **End if**
15.   **End for**
16. **End if**
17. **If** ($Node(e_1)$ and $Node(e_2)$ are not leaf nodes)
18.   **For** (every object $se_1 \in Node(e_1)$ and $se_2 \in Node(e_2)$ in w)
19.   Prune such nodes se_1 and se_2 whose distances to their corresponding root nodes are greater than T;
20.   Compute *MinMaxDist* between se_1 and se_2 denoted as *dist*;
21.   **If** ($dist < \theta$ and there are k elements in *queue*)
22.   Update θ to be the maximum value between the distance value of the top element in *queue* and *dist*;
23.   Compute the *MinMinDist* value between se_1 and se_2;
24.   **If** (it is less than θ)
25.   Push $(se_1, se_2, MinMinDist)$ to *queue*;
26.   **End if**
27.   **End if**
28.   **End for**
29. **End if**
30. **If** (e_1 is a leaf node and e_2 is an internal node)
31.   **For**(every object $se_2 \in Node(e_2)$ in w)
32.   Prune se_2 whose distance to its root node is greater than θ;
33.   **End for**
34. **End if**
35. Use the similar manner as shown in lines 17–29 to handle the case that e_1 is a leaf node and e_2 is an internal node;
36. **End while**
37. Pop all triples from *temp*;

$$r_n = \begin{cases} 1.5^* r_{n-1} & \text{if count} = 0 \\ \sqrt{\dfrac{k}{D(win)}} & \text{if } 0 < \text{count} \leq k-1 \end{cases} \tag{6}$$

where $D(win)$ is the density of the window defined as follows:

$$D(win) = \frac{count}{(win.x_u - win.x_l) \times (win.y_u - win.y_l)} \tag{7}$$

Finally, we use the similar manner as shown in lines 5–36 to find other CPs.

The WinQuery algorithm plays an essential role in coping with *k*-CCPQ problem. The time complexity of WinQuery is similar to that of the Heap algorithm [4] and

SRCP-tree [7]. However, WinQuery performs better than the above two algorithms. This is because it uses a threshold θ as a filter to compress the size of the queue and applies an LRU buffer policy to cache the index entities. We will compare and analyze the performance of the above three algorithms in Section 5.

In WinQuery, we compute the closest pairs based on Equation (8). There is at least one pair (s_i, t_j) where $s_i \in S$, $t_j \in T$, such that:

$$\text{MinMinDist}(S, T) \leq \text{Dist}(s_i, t_j) \leq \text{MinMaxDist}(S, T) \qquad (8)$$

Essentially, the proposed k-CCPQ algorithm can be generalized to handle other k-CPQ problems and is appropriate for other spatial objects. For instance, WindQuery is applicable to achieve continuous monitoring of nearest neighbors in highly dynamic scenarios where the objects move frequently and arbitrarily.

5 Experiments

The proposed algorithms were implemented in Java using spatial index library [15]. In order to measure the performance of GWCCP, we compare it with the typical k-CPQ processing algorithm, i.e., the Heap algorithm [4] that is non-recursive and evaluated to be better for the k-CPQ problem than other algorithms proposed in [4], and SRCP-tree [7]. To facilitate comparison, we extended Heap to handle the k-CCPQ problem in a given query range. Here, we call the new heap-based algorithm RHeap for short.

For each set of experiments, we use the following real-world and synthetic data sets.

- The synthetic data sets of distinct cardinalities are denoted by points with x and y coordinates. They are the sample data for the crime mapping and analysis tool CrimeStat [16] following a Bayesian distribution. They are used in the Journey to Crime module. CrimeStat is a spatial statistics package that can analyze crime incident location data and it provides a variety of tools for the spatial analysis of crime incidents [17].
- Two real-world data sets are from sample programs of Crime Travel Demand Module in CrimeStat. They consist of 65,536 traffic analysis zones represented by longitude and latitude. The data generating involves putting together the necessary data to estimate the model. This includes selecting an appropriate zone system, obtaining data on crime trips and allocating it to zones, obtaining zonal variables that will predict trips, creating possible policy, and obtaining one or more modeling networks [17]. The detailed description is available at http://www.icpsr.umich.edu/CRIMESTAT/files/CrimeStatChapter.11.pdf.

We conduct experiments on a PC of 2.4 GHz Pentium 4 processor with 512 MB of RAM. To comply with RHeap, the tree node capacity was set to 21, and the minimum capacity was set to 7, due to the reason given in [18]. To facilitate comparison, each experiment was run ten times and the average value was used to evaluate the performance.

We perform extensive experiments aiming to compare the performance of GWCCP with RHeap over two distinct data sets, and with SRCP-tree within the same data set, respectively, for the *k*-CCPQ problem in the crime databases. The experimentations consist of evaluating the effect of three important factors, i.e., the portion of overlapping between two data sets, the value of *k*, and the LRU buffer size.

5.1 Query Time Comparison of 1-CCPQ Algorithms

In this section, we compare the query time performance between GWCCP and RHeap. Both algorithms are evaluated with respect to the size of the data sets in distinct portions of overlapping between two data sets. To facilitate comparison, we assume zero buffer size for C-trees in this experiment.

Figure 2 illustrates the performance of both algorithms on 1-CCPQ between synthetic data under varying cardinality in (a) 0% and (b) 100% overlapping workspaces, respectively. The memory cost of 1-CCPQ for real-world data sets is shown in Fig. 3. Notice that the similar comparison results can be found for any other value of *k* as well, here we only give the results when *k*=1.

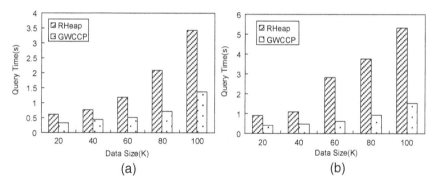

Fig. 2 Query time comparison of 1-CCPQ algorithms on synthetic data in: (**a**) 0% and (**b**) 100% overlapping workspaces

As shown in Fig. 2, GWCCP decreases the query time with respect to RHeap by a factor of 1.8–2.5 for no overlap cases between two data sets. When the data sets overlap (Fig. 2(b)), GWCCP also performs well, and achieves time saving of 1–4 times. One reason is that C-tree uses the shortest distance threshold θ instead of zero as the priority of the closest pairs, which increases the chance that a pair can be pruned from the priority queue in order to reduce the computations of *k*-CCPQs between two distinct data sets. In Fig. 3, the memory cost remains almost unchanged as the data set grows due to the threshold θ that helps save the memory. In addition, we can see that the query time in terms of GWCCP does not increase drastically with the data size.

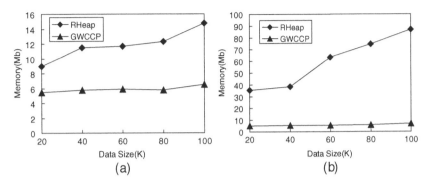

Fig. 3 Memory cost comparison of 1-CCPQ algorithms on real-world data in: (**a**) 0% and (**b**) 100% overlapping workspaces

We conclude that GWCCP is a better solution for the k-CCPQ problem, especially when the data sets significantly overlap. Since both algorithms are sensitive to the overlap factor, we have to further discuss the effect of this parameter in the following section.

5.2 The Effect of Overlap

We compare the memory cost of GWCCP with that of RHeap in the synthetic and the real-world data sets of 40 K cardinality as the overlapping percentage changes from 0 to 100%. The results are shown in Fig. 4.

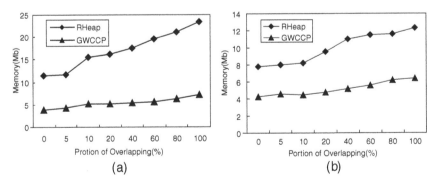

Fig. 4 Memory cost comparison on the overlap factor with: (**a**) synthetic and (**b**) real-world data

Straightforwardly, the overlap between two data sets plays a critical role for the performance. However, for GWCCP, the memory cost for a query involving larger overlap is slightly higher than the case involving disjointed workspaces. Because the query window extended gradually until finding the k-CPQs, and the growing query window helps eliminate the unneccessary memory cost spent on computing the distances between spatial objects. For RHeap, the higher the portion of overlapping between two data sets, the higher the memory cost. In summary, GWCCP is more suitable for *k*-CCPQ with a higher portion of overlapping between two distinct data sets.

5.3 The Effect of k

For this set of experiments, we run *k*-CCPQs in the real-world and in the synthetic data sets of 40 K cardinality, with *k* varying from 1 up to 100,000. Figures 5 and 6 illustrate the query time and the memory cost of both algorithms assuming (a) 0% and (b) 100% overlapping workspaces in the real-world data sets.

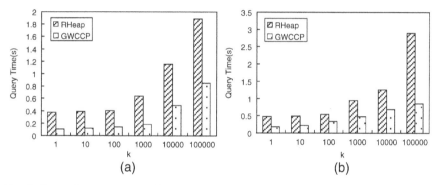

Fig. 5 Query time comparison of *k*-CCPQ algorithms with real-world data in: (**a**) 0% and (**b**) 100% overlapping workspaces

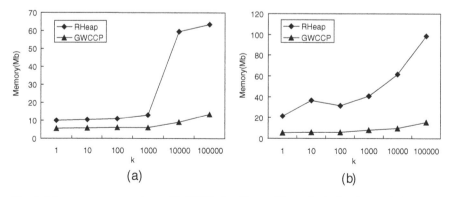

Fig. 6 Memory cost comparison of *k*-CCPQ algorithms with synthetic data in: (**a**) 0% and (**b**) 100% overlapping workspaces

According to Fig. 5, the cost of both algorithms increases with k in the real-world data sets. This is because both algorithms need more time to find the increased closest pairs of spatial objects as k grows. GWCCP wins RHeap with an average gap of 65% and 53% in 0% and 100% overlapping workspaces, respectively.

We can see from Fig. 6 that GWCCP achieves a significant improvement in terms of memory cost by a factor of 2–5 for 0% overlap (respectively, 4–7 for 100% overlap) as k grows. This further illustrates that GWCCP adapts to cope with the case of higher overlap between two data sets. According to Fig. 6, when the k value is larger than 1,000, the memory cost of RHeap changes sharply. However, GWCCP has a slight change when k increases. This is because C-tree uses the LRU buffer policy [13] to improve the quality of the C-tree update operation. In the following section, we will further investigate the effect of the LRU buffer.

5.4 Disk Access Comparison Under Distinct Buffer Size

Buffer policies considerably affect the performance of R-tree [19]. In this set of experiments, the buffer varies from $B = 0, \ldots, 256$ pages, i.e., each C-tree has equal portions of $B/2$ pages. We observe the performance of each algorithm in the real-world as well as the synthetic data sets of 40 K cardinality assuming 20% overlap and $k = 1000$. It is a tradeoff to choose a relatively low overlapping percentage, and RHeap performs well when k is lower than 1000 as empirically illustrated in the previous subsection. The results are shown in Fig. 7.

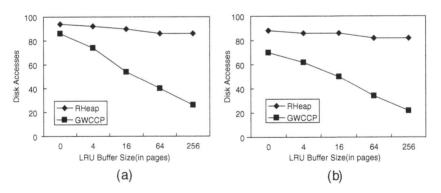

Fig. 7 Comparison of k-CCPQ algorithms under distinct LRU buffer size in: (**a**) synthetic and (**b**) real-world data sets

Figure 7 shows that the results for the synthetic and the real-world data sets are improved by up to 3 times in terms of disk accesses. However, RHeap is not sensitive to the buffer size (only up to 10% improvement as the buffer size reaches 256). On the contrary, GWCCP is sensitive to the buffer size. Because GWCCP uses the LRU buffer policy to maintain the C-trees, the disk accesses are greatly reduced.

5.5 *Comparison Between GWCCP and SRCP-Tree*

The problem of finding *k*-CCPQ in the same data sets is another practical problem in real-world scenarios. For example, a police officer may want to find the two closest first-aid centers. As suggested by [7], SRCP-tree performs well in handling this problem. In this section, we compare the performance in terms of the query time and memory cost of GWCCP with SRCP-tree.

In this set of experiments, we compare these two algorithms with an 80 K real-world and 80 K synthetic data sets. We first observe the query time of GWCCP and SRCP-tree as the overlapping percentage ranges from 0 to 100% with an interval of 20%. The results are shown in Fig. 8.

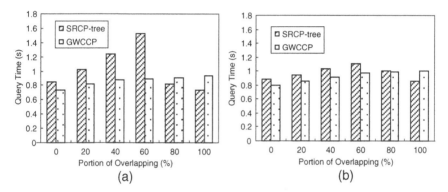

Fig. 8 Query time comparison within the same data set under distinct portions of overlapping on: (**a**) synthetic and (**b**) real-world data

As shown in Fig. 8, GWCCP outperforms SRCP-tree when the portion of overlapping is lower than 60% in the real-world and the synthetic data sets. For GWCCP, the query window is gradually extended until it finds the *k*-CCPQs (the overlapping percentage is about 60%), which helps eliminate the re-calculation of CPs. In particular, the increase of query time in terms of GWCCP is slight as it finds the *k*-CCPQs, whereas the query time in terms of SRCP-tree decreases drastically when the overlapping percentage is higher than 60%. This is because the probability of having a CP in the query range increases as it becomes large, and thus the query time drops.

We also compare the memory cost of these two methods within the synthetic and the real-world data sets, the results are shown in Fig. 9.

As we can see from Fig. 9, the memory cost of GWCCP keeps flat when the portion of overlapping reaches 60%, whereas the curve of SRCP-tree goes up linearly with the overlapping percentage. The reason is that GWCCP will stop searching when it finds the *k*-CCPQs, even if the portion of overlapping increases.

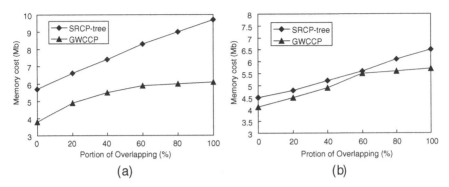

Fig. 9 Memory cost within the same data set under distinct portions of overlapping on: (**a**) synthetic and (**b**) real-world data

6 Conclusions and Future Work

We have proposed a new constrained k-closest pairs query processing algorithm based on growing windows, namely GWCCP. It employs an R-tree structure having inherent properties of the SRCP-tree to store the index entities, and a density-based range estimation method without boundary to calculate the square query range. Experiments have demonstrated that GWCCP outperforms the heap-based algorithm with respect to the portion of overlapping between two distinct data sets, the value of k, and the LRU buffer size.

We point out several interesting issues for future research including developing other range estimation approaches for k-CCPQs, proposing other good buffer policy for C-trees, and applying our proposed algorithm to assist in crime mapping and data analysis.

Acknowledgments This work is supported by the National Natural Science Foundation of China under Grant No. 60773169, the 11th Five Years Key Programs for Sci. and Tech. Development of China under Grant No. 2006BAI05A01, and the Youth Software Innovation Project of Sichuan Province under Grant Nos. 2007AA0032 and 2007AA0028.

References

1. Overview of Data Collection in British Columbia. Available at http://www.pssg.gov.bc. ca/police_services/publications/
2. Chen H (2007) Exploring extremism and terrorism on the Web: the dark web project. In: Pacific Asia Workshop on Intelligence and Security Informatics, PAISI 2007, Chengdu, pp. 1–20
3. Leipnik MR, Albert DP (2003) GIS in law enforcement: Implementation issues and case studies. Routledge, London, pp. 1–47
4. Corral A, Manolopoulos Y, Theodoridis Y, Vassilakopoulos M (2000) Closest pair queries in spatial databases. In: Proceedings of ACM SIGMOD 2000, Dallas, pp. 189–200

5. Corral A, Manolopoulos Y, Theodoridis Y, Vassilakopoulos M (2004) Algorithms for processing *k*-closest pair queries in spatial databases. Data and Knowledge Engineering, 49(1): 67–104
6. Qiao S, Tang C, Jin H, Dai S, Chen X (2008) Constrained *k*-Closest Pairs Query Processing Based on Growing Window in Crime Databases. In: 2008 IEEE International Conference on Intelligence and Security Informatics, Taipei, pp. 58–63
7. Shan J, Zhang D, Salzberg B (2003) On spatial-range closest-pair query. In: Proceedings of SSTD 2003, Greece, pp. 252–270
8. Guttman A (1984) R-trees a dynamic index structure for spatial searching. In: Proceedings of ACM SIGMOD 1984, Boston, pp. 47–57
9. Roussopoulos N, Kelley S, Vincent F (1995) Nearest neighbor queries. In: Proceedings of ACM SIGMOD 1995, San Jose, pp. 71–79
10. Liu D, Lim E, Ng W (2002) Efficient k nearest neighbor queries on remote spatial databases using range estimation. In: Proceedings of SSDBM 2002, Edinburgh, pp. 121–130
11. Hjaltason GR, Samet H (1998) Incremental distance join algorithms for spatial databases. In: Proceedings of ACM SIGMOD 1998, Seattle, pp. 237–248
12. Ferhatosmanoglu H, Stanoi I, Agrawal D, El Abbadi A (2001) Constrained Nearest Neighbor Queries. In: Proceedings of SSTD 2001, Redondo Beach, pp. 257–278
13. Bhide A, Dan A, Dias D (1993) A simple analysis of LRU buffer replacement policy and its relationship to buffer warm-up transient. In: Proceedings of ICDE 1993, Vienna, pp. 125–133
14. Hsiao P, Tsai C (1990) A new plane-sweep algorithm based on spatial data structure for overlapped rectangles in 2-D plane. In: COMPSAC'90, Chicago, pp. 347–352
15. http://research.att.com/~marioh/spatialindex/index.html
16. http://www.icpsr.umich.edu/CRIMESTAT/
17. Ned Levine (2007) CrimeStat: A Spatial Statistics Program for the Analysis of Crime Incident Locations (v 3.1). Ned Levine and Associates, Houston, TX, and the National Institute of Justice, Washington, DC. March.
18. Beckmann N, Kriegel HP, Schneider R, Seeger B (1990) The R*-tree: an efficient and robust access method for points and rectangles. In: Proceedings of ACM SIGMOD 1990, New York, pp. 322–331.
19. Leutenegger ST, Lopez MA (2000) The effect of buffering on the performance of R-trees. IEEE Transactions on Knowledge and Data Engineering, 12(1): 33–44.

What-If Emergency Response Through Higher Order Voronoi Diagrams

Ickjai Lee, Reece Pershouse, Peter Phillips, Kyungmi Lee, and Christopher Torpelund-Bruin

Abstract As witnessed in many recent disastrous events, what-if emergency response is becoming more and more important to prevent hazards, to plan for actions, to quickly respond to minimize losses, and to recover from damages. In this chapter, we propose a complete set of higher order Voronoi diagram-based emergency response system for what-if analysis which is particularly useful in highly dynamic environments. This system is based on a unified order-k Delaunay triangle data structure which supports various topological and regional queries, and what-if analysis. The proposed system encompasses (1) what-if scenarios when new changes are dynamically updated; (2) what-if scenarios when order-k generators (disasters or professional bodies) or their territorial regions are of interest; (3) what-if scenarios when ordered order-k generators or their territorial regions are of interest; (4) what-if scenarios when k-th nearest generators or their territorial regions are of interest; and (5) what-if scenarios with mixtures of the above.

Keywords Higher order Voronoi diagrams · Emergency management · Generalized Voronoi diagrams · What-if analysis

1 Introduction

Natural and man-made disasters are constantly occurring in our daily life, and they are greatly affecting securities of individual, regional, national, and international levels. These disastrous events are detrimental to people, property, the environment, and homeland security. Lack of appropriate emergency management can lead to environmental, financial, and structural damages, losses, or destruction. Even if it is almost impossible to avoid occurrences of disasters, its prediction and

Ickjai Lee, Reece Pershouse, Peter Phillips, Kyungmi Lee, and Christopher Torpelund-Bruin
School of Business (IT), James Cook University, Cairns, QLD 4870, Australia
e-mail: ickjai.lee@jcu.edu.au; reece.pershouse@jcu.edu.au;
peter.phillips@jcu.edu.au; joanne.lee@jcu.edu.au;
christopher.torpelund@jcu.edu.au

C.C. Yang et al. (eds.), *Security Informatics*, Annals of Information Systems 9,
DOI 10.1007/978-1-4419-1325-8_5, © Springer Science+Business Media, LLC 2010

preparedness along with an effective post-emergency response program can mitigate the risk and damages [1]. A general-purpose emergency response system providing what-if decision support information for every stage of emergency management (mitigation, preparedness, response, and recovery) for diverse emergency scenarios is in great demand [2–4].

Geographic Information Systems (GIS) have been one of the popular tools and have been widely used in various aspects of emergency response and management [5–10]. They support features such as data capturing, structuring, preprocessing, mapping, visualization, and also zonal and focal analysis. However, most of them are limited to producing cartographic mappings, basic simulations, and visualization rather than emergency planning, preparedness, what-if response, and predictive modeling [11]. One of the intrinsic properties of disastrous activities is of their inherent dynamic nature. They are active and unpredictable. What-if analysis for various scenarios is of great importance to the understanding of their dynamics and to emergency planning. For instance, in a situation where several emergency response units are required to collaborate to recover damaged areas to minimize losses when a disaster occurs, government agencies would be interested in which area is covered by a certain set of emergency professional bodies, how many bodies must be involved in the recovery, what happens when the first few nearest bodies are engaged in other disaster recoveries, and so forth. Current GIS provide limited functions enabling what-if analysis, and thus computer software simulating various scenarios for what-if analysis is in great demand [11, 12]. There exist several emergency management simulators [13–16]. They provide tools to model movement and behavior of people, and help people systematically respond when a disaster occurs. However, they fail to provide a general purpose what-if emergency response toolbox handling various scenarios for timely and well-informed decision makings in highly dynamic environments.

What-if analysis is exploratory and interactive [17]. It explores scenarios as soon as changes are suggested. Thus, a computer system providing what-if analysis is typically equipped with a data structure supporting fast local updates and consistent spatial tessellations. The Voronoi diagram and its dual Delaunay triangulation provide a robust framework for exploring various scenarios for what-if analysis, and have been widely applied to many aspects of geospatial analysis [18]. The Voronoi diagram provides consistent natural neighboring information overcoming the inconsistent modeling of traditional raster and vector models [19]. Its dual triangulation provides a robust data structure framework for dynamic local updates. In addition, there exist many generalizations on the Voronoi diagram supporting what-if analysis. Several attempts [12, 17] have been made to employ the flexibility (generalizability) of Voronoi tessellation for what-if analysis. However, they are limited to certain environments requiring extreme and confined generalizations, and as such are not general enough to cover various scenarios for managing emergencies in highly dynamic environments.

In this chapter, we propose a generalized Voronoi diagram-based emergency response and management system for what-if analysis particularly useful in highly dynamic environments. It builds a unified order-k Delaunay triangle data structure from which users can derive the complete set of order-k Voronoi diagrams, the complete set of ordered order-k Voronoi diagrams, and the complete set of k-th nearest Voronoi diagrams. Note that emergency planning and recovery are based on the scale of the disaster. These generalized Voronoi diagrams provide useful information to situations where more than k emergency response units are required to get involved in a recovery. The proposed system supports dynamic updates (insertion and deletion of new disasters or emergency response units) and various geometrical and topological neighboring information including ordered/unordered k-nearest neighbors. In addition, the proposed what-if emergency response system prop up the four essential types of what-if queries in emergency response systems: (1) what-if neighboring queries; (2) what-if zoning (districting) queries; (3) what-if facility locating queries; and (4) what-if routing queries.

The rest of the chapter is organized as follows. Section 2 defines the higher order Voronoi diagrams and k-th nearest Voronoi diagram. It also investigates their interrelationships and algorithmic procedures. Section 3 discusses the working principle of our proposed system. It provides an algorithmic procedure of Voronoi-based what-if emergency management system, and explores its capabilities. Section 4 explains various what-if analysis approaches. Section 5 provides a number of scenarios to which our proposed system can be applied. Section 6 illustrates an extension into Minkowski metrics, whilst Section 7 draws concluding remarks.

2 What-If Modeling with Higher Order Voronoi Diagrams

The ordinary Voronoi diagram of a set $P = \{p_1, p_2, \ldots, p_n\}$ of generators in a study region S tessellates it into mutually exclusive and collectively exhaustive regions. Each region has a generator closest to it. This geospatial tessellation provides natural neighbor relations that are crucial for many topological queries in geospatial modeling and analysis, whilst its dual graph, the Delaunay triangulation, provides a robust framework for structural arrangements of the Voronoi diagram. This geospatial tessellation has many generalizations and this flexibility provides a robust framework for what-if and what-happens modeling and analysis [12]. Higher order Voronoi diagrams are natural and useful generalizations of the ordinary Voronoi diagram for more than one generator [18]. They provide tessellations where each region has the same k (ordered or unordered) closest sites for a given k. These tessellations are useful for situations where more than one location of interest are not functioning properly (engaged, busy, closed or fully scheduled) or several locations are required to work together.

The order-k Voronoi diagram $v^{(k)}$ is a set of all order-k Voronoi regions $v^{(k)} = \left\{ V\left(P_1^{(k)}\right), \ldots, V\left(P_n^{(k)}\right) \right\}$, where the order-$k$ Voronoi region $V\left(P_i^{(k)}\right)$ for a random subset $P_i^{(k)}$ consisting of k points out of P is defined as follows:

$$V\left(P_i^{(k)}\right) = \left\{ p \mid \max_{p_r \in P_i^{(k)}} d\left(p, p_r\right) \leq \min_{p_s \in P \setminus P_i^{(k)}} d\left(p, p_s\right) \right\}. \tag{1}$$

In $v^{(k)}$, k generators are not ordered, however, in some emergency management situations an ordered set would be of interest. This ordered set can be modeled by the ordered order-k Voronoi diagram $v^{<k>}$. It is defined as $v^{<k>} = \{V(P_1^{<k>}), \ldots, V(P_m^{<k>})\}$, where $m = n(n-1)\ldots(n-k+1)$, and the ordered order-k Voronoi region $V(P_i^{<k>})$ is defined as

$$V\left(P_i^{<k>}\right) = \left\{p \mid d(p, p_{i1}) \leq \ldots \leq d(p, p_{ik}) \leq d(p, p_j), p_j \in P \setminus \{p_{i1}, \ldots p_{ik}\}\right\}.$$
(2)

$V(P_i^{<k>})$ is a refinement of $V\left(P_i^{(k)}\right)$, namely $V\left(P_i^{(k)}\right) = \cup_{P_j^{<k>} \in A^{<k>}}\left(P_i^{(k)}\right)$

$V\left(P_j^{<k>}\right)$, where $A^{<k>} = \left(P_i^{(k)}\right)$ is the set of all possible k-tuples made of $p_{i1}, \ldots p_{ik}$ [18].

One generalized variant of the ordinary Voronoi diagram similar to $v^{(k)}$ is the k-th nearest Voronoi diagram. This is particularly useful when users are interested in only k-th nearest region. The k-th nearest Voronoi diagram $v^{[k]}$ is a set of all k-th nearest Voronoi regions $v^{[k]} = \{V^{[k]}(p_1), \ldots V^{[k]}(p_n)\}$, where the k-th nearest Voronoi region $V^{[k]}(p_i)$ is defined as

$$V^{[k]}(p_i) = \left\{p \mid d(p, p_i) \leq d(p, p_j), p_j \in P \setminus \{k \text{ nearest points to } p_i\}\right\}.$$
(3)

$V(P_i^{<k>})$ is a refinement of $V\left(P_i^{[k]}\right)$. That is, $V\left(P_i^{[k]}\right) = \cup_{(p_{j1}, \ldots p_{jk-1}) \in A^{<k-1>}(P \setminus \{p_i\})}$

$V\left((p_{j1}, \ldots p_{jk-1}, p_i)\right)$, where $A^{<k-1>}(P \setminus \{p_i\})$ is the set of all possible $(k-1)$-tuples consisting of $k-1$ elements out of $P \setminus \{p_i\}$ [12, 17, 18].

Figure 1 illustrates the relationship of various higher order Voronoi diagrams with a set P of five generators, $\{p_3, p_4, p_5, p_6, p_7\}$. Figure 1(a) shows the ordinary Voronoi diagram of P. The shaded Voronoi region in Fig. 1(a) is $V(p_4)$ having p_4 as the closest generator. Figure 1(b) depicts the order-3 Voronoi diagram of P, $v^{(3)}(P)$. The shaded region $V\left(P_{\{p_4, p_6, p_7\}}^3\right)$ has $\{p_4, p_6, p_7\}$ as the first three closest generators. In this diagram, generators are not ordered. Figure 1(c) shows the ordered order-3 Voronoi diagram of P, $v^{<3>}(P)$. Here, the shaded region in Figure 1(b) is further split into six exclusive ordered order-3 Voronoi regions. Figure 1(d) depicts the 3rd-nearest Voronoi diagram of P, $v^{[3]}(P)$. The shaded order-3 Voronoi region in Fig. 1(b) is further decomposed into three 3rd-nearest Voronoi regions.

Several algorithmic approaches [20–23] have been proposed to compute higher order diagrams. Dehne [22] proposed an $O(n^4)$ time algorithm that constructs the complete $v^{(k)}$. Several other attempts [20, 21, 23] have been made to improve the computational time requirement of Dehne's algorithm without explicit underlying data structure. The best known algorithm for the order-k Voronoi diagram is $O\left(k(n-k)\log n + n\log^3 n\right)$ [24]. Thus, the complete set of $v^{(k)}$ requires $O(n^3 \log n)$ time. Our system builds a unified Delaunay triangle data structure that enables us to derive the complete set of $v^{(k)}$, $v^{<k>}$ and $v^{[k]}$ for various what-if

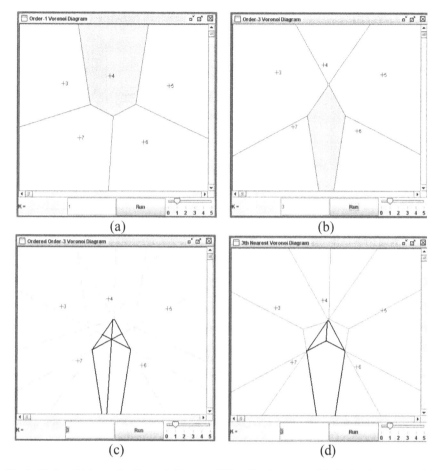

Fig. 1 Various higher order Voronoi diagrams ($|P| = 5$) where $p_i \in P$ is depicted as a cross (+) and an associated identification number i: (**a**) $v^{(1)}(P)$; (**b**) $v^{(3)}(P)$; (**c**) $v^{<3>}(P)$; (**d**) $v^{[3]}(P)$

analyses. The Delaunay triangle-based data structure well supports various geospatial topological queries including natural neighbor and region queries. We implement $v^{(k)}$ based on Dehne's algorithm and extend it to include the complete set of $v^{<k>}$ based on $v^{<k>} = v^{<k-1>} \cup v^{(k)}$, for $k = 2, \ldots, n - 1$. In addition, we extend to include the complete set of $v^{[k]}$ based on $v^{[k]} = v^{(k-1)} \cup v^{(k)}$, for $k = 2, \ldots, n-1$. Note that when $k = 1$, $v^{(1)} = v^{<1>} = v^{[1]}$.

3 What-If Emergency Response Through Higher Order Voronoi Diagrams

3.1 Framework of the Proposed What-If System

What-if system is required to be dynamic and simple to use for users that may not be domain experts. In Fig. 2, our proposed system has three basic components:

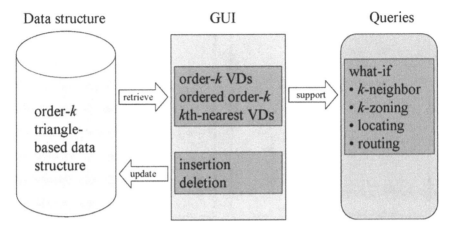

Fig. 2 Framework of the proposed system

data structure, Graphical User Interface (GUI), and what-if queries. The data structure implements an order-k triangle-based data structure that will be discussed in the next subsection. The data structure needs to be dynamic and unified, to provide interactive and efficient what-if analysis of different visualization and every changing data. The GUI interface has been designed to support a number of the functions needed for exploratory spatial analysis [25], including re-expression, dynamic updates (insertion and deletion) of multiple views, linked views, and basic queries. Multiple windows allow different visualizations simultaneously. Linked Windows were used to allow the user to see the selected region for a selected position for both different orders and types of diagrams, simultaneously. Finally re-expression allows the user to see different views of the same spatial data, including $v^{(k)}$, $v^{<k>}$, and $v^{[k]}$. It also allows the user to view the underlying data structure. For real data set, it supports Cartesian point with large double values and allows data to be bulk loaded. The save and load function was also included to deal with common separated value file (common file format with real data sets) and also object file to store the actual data structure for quicker load time for previously generated data. Therefore, it is useful for both what-if analysis and loading real data set.

3.2 Complete Order-k Delaunay Triangle Data Structure

A flexible data structure supporting multiple models for diverse what-if scenarios is a must for highly dynamic emergency management systems. The data structure needs to efficiently support rapid transits from model to model, and effectively manage various models. Our proposed system is based on a unified triangle-based data structure that robustly supports all $v^{(k)}$, $v^{<k>}$, and $v^{[k]}$. The unified data structure stores all order-k Delaunay triangles to efficiently support the complete $v^{(k)}$, $v^{<k>}$,

Table 1 Complete order-k Delaunay triangle data structure

Order-k Delaunay triangle Δ	Inpoint
Order-0 Delaunay triangle $\Delta p_a p_b p_c$	ϕ
...	ϕ
Order-1 Delaunay triangle $\Delta p_i p_j p_k$	$p \in P$
...	...
...	...
Order-$(n-3)$ Delaunay triangle $\Delta p_q p_r p_s$	$P \setminus \{p_q p_r p_s\}$
...	...

and $v^{[k]}$ spontaneously. Table 1 depicts the complete order-k Delaunay triangle data structure.

It stores the complete order-k Delaunay triangles and consists of two fields: order-k triangle and inpoints. Here, a triangle $\Delta p_l p_m p_n$ becomes an order-k triangle if its circumcircle contains k other points (inpoints) within it. For instance, the lowest degree triangles, order-0 Delaunay triangles, do not contain any other points within their circumcircles whilst the highest degree triangles, order-$(n$–$3)$ triangles, contain all points in P except three points forming the triangles. An algorithmic procedure of our system based on this data structure is described in the following subsection.

3.3 Algorithm and Analysis

Our system takes a set P of generators as an input and builds a unified order-k Delaunay triangle data structure from which the complete set of $v^{(k)}$, $v^{<k>}$, and $v^{[k]}$ can be derived. The pseudocode of our proposed system is as follows:

Algorithm Voronoi-based What-if Emergency Response System

Input: A set $P = \{p_1, p_2, \ldots, p_n\}$ of generators;
Output: Unified order-k Delaunay triangle data structure (DS), and
       complete set of $v^{(k)}$, $v^{<k>}$, and $v^{[k]}$;
1) **begin**
2) $DS \Leftarrow BulkLoad\ (P)$;
3) $DrawDiagrams\ (DS)$;
4) **do**
5) $\begin{cases} DS \Leftarrow DynamicUpdate\ (p)\ ; \\ Retrieve\ (p)\ ; \end{cases}$
6) $DrawDiagrams\ (DS)$;
7) **while** the user exits;
8) $Save\ (DS)$;
9) **end**

Initially, *BulkLoad* (P) loads a data set and builds *DS* to begin with. This step builds an entire set of order-k Delaunay triangles as a unified data structure for the complete set of $v^{(k)}$, $v^{<k>}$, and $v^{[k]}$. Since every possible triangle $\Delta p_i p_j p_k$ needs to be tested for which order-k triangle it belongs to, this step requires $O\left(n^4\right)$ time. In the next step, *DrawDiagrams* (DS) derive and draw $v^{(k)}$, $v^{<k>}$ (for $1 \leq k \leq n-1$), and $v^{[k]}$ (for $1 \leq k \leq n$) at the user's choice of k. Note that, multiple k can be chosen in multiple windows. This step implements [22] for the derivation of the complete set of $v^{(k)}$ from *DS*. Once $v^{(k)}$ is computed, $v^{<k>}$ is derived from $v^{<k-1>} \cup v^{(k)}$, for $k = 2, \ldots, n-1$, whilst $v^{[k]}$ is derived from $v^{(k-1)} \cup v^{(k)}$, for $k = 2, \ldots, n-1$.

In Step 5, users can have two different modes: edit and retrieval. In edit mode (*DynamicUpdate* (p)), users can either add a new generator p or remove it from P. This update consists of two substeps. First it needs to go through every order-k triangle in *DS* to check if its circumcircle contains p or not. If it does, then it is marked as an order-$(k+1)$ triangle, otherwise it is left unchanged. Second, the dynamic update step will create new order-k triangles $\Delta p_i p_j p$ with $p_i p_j \in P$. Deletion is implemented in a similar way. Both substeps require $O\left(n^3\right)$, thus *DynamicUpdate* (p) requires cubic time. In retrieval mode, users are able to retrieve topological information for a given location p (mouse click) from the complete set of $v^{(k)}$, $v^{<k>}$, and $v^{[k]}$. Retrievable topological information includes ordered and unordered k-nearest neighbors to p, Voronoi regions ($v^{(k)}$ (p), $v^{<k>}$ (p), and $v^{[k]}$ (p)) and their perimeters and areas, and their topologically adjacent Voronoi neighbors. Step 6 requires $O\left(n^3\right)$ time, thus our proposed system requires $O\left(n^4\right)$ time.

3.4 Supporting What-If Queries

In emergency management, what-if queries for diverse emergency decision makings are of importance. It is crucial for residents to find the k-th nearest evacuation center from their houses to help them plan for evacuation. It is vital for emergency responding units to find the region in which the k-th nearest emergency unit is the emergency unit at a certain location. Also, it is essential to locate emergency responding units equally reachable from neighboring disasters. In addition, it is important to find safe road networks from residential places to the k-th nearest evacuation shelter. Our model supports these questions and its capability encompasses:

- *Type I*: k-neighboring query
  Topological neighborhood (adjacency) is a key factor in emergency management systems. Geospatial arrangement of emergency units (or disasters) greatly affects planning and making decisions. Type I supports k-neighboring queries in diverse cases. The complete set of $v^{(k)}$, $v^{<k>}$, and $v^{[k]}$ returns unordered order-k, ordered order-k, and k-th nearest neighbors for all k. As our underlying order-k Delaunay triangle data structure supports the complete set of $v^{(k)}$, $v^{<k>}$, and $v^{[k]}$, this type of queries can be easily realized by our proposed framework. Note that, Delaunay triangles are explicitly coding topological neighboring relations.

In retrieval mode in our system, k-neighboring information can be extracted by clicking a mouse within S.

- *Type II*: districting query

Districting segments the study region into meaningful and manageable tessellations in which each disaster influences or of which each emergency unit is in charge. It is important for preparedness, response, and recovery. Various higher order Voronoi regions explicitly provide districting and zoning information. Figure 3(a) is a screen capture of our system highlighting an order-2 Voronoi region and its two nearest neighbors when a mouse was clicked. Thus, it clearly demonstrates that our system interactively supports the first two types of queries.

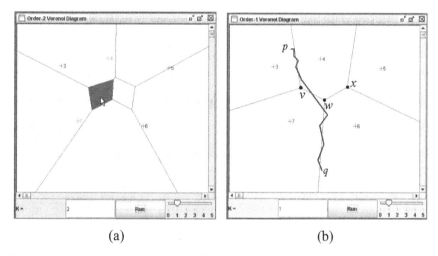

Fig. 3 Queries support with higher order Voronoi diagrams (P is the same as in Fig. 1): (a) $v^{(1)}$ (P); (b) $v^{(1)}$ (P) with two locations p and q, three Voronoi vertices v, w, x, and a road network straddling Voronoi edges

- *Type III*: location optimization support

Responding to and recovering from a disaster with limited resources are not easy tasks. This type supports locating multiple emergency units in right places when multiple disasters occur in order to mitigate damages. Since Voronoi vertices are locations equi-distant from neighboring generators (disasters), higher order Voronoi vertices are indicative of this type of queries. This can be manifested in Fig. 3(b). When an emergency response body is planning to locate two regional emergency response units in the study region to handle diasters P, then Voronoi vertices v and x could be a good option since v can handle three disasters (p_3, p_4, p_7) whilst x can handle the remaining two neighboring disasters (p_5, p_6). Also, these two v and x are equi-distant from their neighboring disasters so that they are not biased to any disaster but rather well-located. In addition, a higher order Voronoi region is an area of ordered or unordered k-nearest neighbors, and it can serve for optimizing locations. For instance, the highlighted region in

Fig. 3(a) is indicative of a possible location of regional emergency unit if it is required to handle two disasters p_4 and p_7 since the region has these two diasters as the first two nearest.

- *Type IV*: routing support

  Finding safe road networks from a given place to the nearest shelter when a disaster occurs is important to both emergency personnel and residents. This supports finding actual road networks as far as possible from disasters indicating less affected pathways to the target shelter. Since Voronoi edges are lines maximizing distance from two neighboring generators (disasters), higher order Voronoi edges can be used for this type of queries. An example is illustrated in Fig. 3(b) with two locations p and q. When disasters P are present, assume that you are required to evacuate from your current location p to an evacuation gathering point q. In this case, you would like to take a route as safe as possible maximizing distance from dangerous disasters. This can be achieved by finding paths that straddle Voronoi edges as shown in blue thick lines in Fig. 3(b).

3.5 What-If Emergency Response Analysis with Dynamic Updates

Dynamic updates (Step 5 of the algorithm in Section 3.3) allow users to explore what-if scenarios to help them make prompt decisions in constantly changing environments. Figure 4 depicts $v^{[2]}$ when P varies from $\{p_3, p_4\}$ to $\{p_3, p_4, p_5, p_6, p_7\}$.

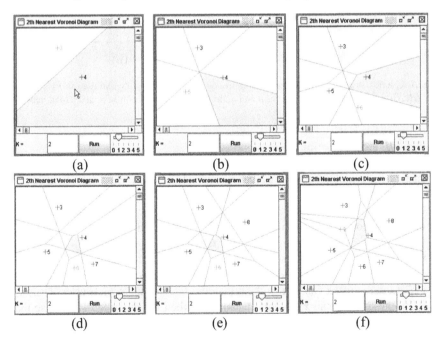

Fig. 4 2nd Nearest Voronoi diagrams with dynamic updates: $2 \leq |P| \leq 7$

Shaded Voronoi regions are highlighted regions when a location is mouse clicked (shown in Fig. 4(a)) in retrieval mode. Also, their corresponding 2nd nearest generators are highlighted in green. A series of dynamic updates shown in Fig. 4 is particularly useful for planning in highly active environments as the interactivity allows the planner to visually inspect and make decisions based on an easily digested representation. Let us consider an evacuation management system, and assume P represents a set of evacuation places and the clicked location is of the user's interest. Figure 4 shows the second nearest evacuation place and its territories from the clicked location as the number of evacuation places vary. This is of particular interest when the first evacuation place is fully booked or roads to it are not accessible. Hence we can determine and visualize very quickly alternative evacuation plans as our environmental factors come into play.

3.6 What-If Emergency Response Analysis with Homogeneous Voronoi Diagrams

This subsection describes emergency management with homogeneous Voronoi diagrams. That is, one type of Voronoi diagram is in use (e.g., order-k Voronoi diagrams in Fig. 5). In many emergency management situations, several professional bodies

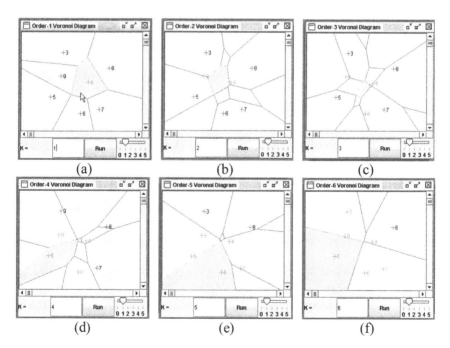

Fig. 5 The complete set of order-k Voronoi diagrams with $|P| = 7$: (**a**) $v^{(1)}$; (**b**) $v^{(2)}$; (**c**) $v^{(3)}$; (**d**) $v^{(4)}$; (**e**) $v^{(5)}$; (**f**) $v^{(6)}$

are required to cooperate to aid with recovery and restoration. This is particularly the case when disasters are severe, meaning rapid action is essential to minimize the damage. For example, many different fire departments may need to work together to extinguish a large forest fire. Figure 5 shows the complete $v^{(k)}$ with the same P as in Fig. 4(f). Highlighted points and regions are order-k neighbors and their corresponding Voronoi regions when the same point as in Fig. 4(a) is clicked. If we assume that a forest fire starts in the clicked location, then this figure provides answers to questions such as which k professional bodies must be involved in the recovery or what are their Voronoi regions. In this particular example, the order-3 Voronoi region of the clicked location gives the smallest area.

3.7 What-If Emergency Response Analysis with Heterogeneous Voronoi Diagrams

This subsection describes emergency management with heterogeneous Voronoi diagrams. That is, different types of Voronoi diagrams are in use simultaneously (e.g., order-k Voronoi diagrams, ordered order-k Voronoi diagrams, and k-th nearest Voronoi diagrams in Fig. 6). Emergency management systems must be able to handle complex and unpredictable environments. A combination of dynamic updates

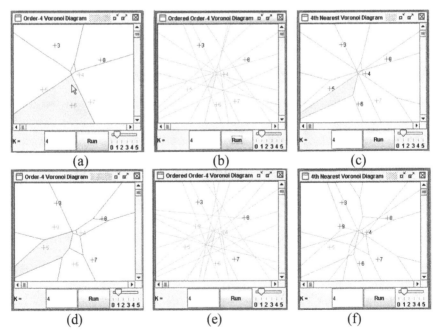

Fig. 6 $v^{(4)}$, $v^{<4>}$, and $v^{[4]}$ of $P = \{p_3, \ldots p_8\}$ and those of $\hat{P} = P \cup \{p_9\}$: (a) $v\left(P^{(4)}\right)$; (b) $v\left(P^{<4>}\right)$; (c) $v\left(P^{[4]}\right)$; (d) $v\left(\hat{P}^{(4)}\right)$; (e) $v\left(\hat{P}^{<4>}\right)$; (f) $v\left(\hat{P}^{[4]}\right)$

and order-k Voronoi diagrams, the ordered order-k Voronoi diagrams and the k-th nearest Voronoi diagrams can be used to explore these complex what-if scenarios. Figure 6 depicts such scenarios.

4 What-If Emergency Response with Real Data Sets

This section examines various $v^{(k)}$, $v^{<k>}$, and $v^{[k]}$ of real data sets from 217 urban suburbs of Brisbane, the capital city of Queensland, Australia. The study region is highly dynamic and active. It continues to experience significant and sustained population growth and various criminal activities [26]. Figure 7 shows a data set within the study region and various Voronoi diagrams. Figure 7(a) depicts 25 murder

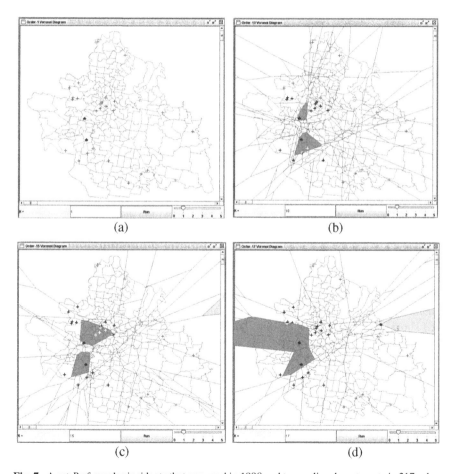

(a)   (b)

(c)   (d)

Fig. 7 A set P of murder incidents that occurred in 1998 and two police departments in 217 urban suburbs of Brisbane, Australia ($|P| = 25$): (**a**) The study region with P and two police departments; (**b**) $v\left(P^{(10)}\right)$; (**c**) $v\left(P^{(15)}\right)$; (**d**) $v\left(P^{(17)}\right)$

incidents (red crosses) that occurred in the study region in the year of 1998 with two police departments (blue houses).

4.1 Case Study 1

This case study models police departments and locations of crimes in Brisbane. Each police department is involved in the same number of (k) of nearest crime investigations in this case study. That is, each police department has the same weight. The study region suffers from various types of crimes and the local government is planning to introduce another police department in the study region for systematic management of crimes. The main aim is to cover all the crimes with the introduction of an additional department within the study region while minimizing k. Manual inspection of this type of investigation is labor intensive. Our framework systematically supports this type of query.

This scenario is of particular interest when crime incidents are excessive to deal with, and each police station has limited resources (policemen and patrol cars etc.), and thus are unable to handle all existing crime incidents. The complete order-k $v^{(k)}$ provides a solution to this scenario. Users need to explore the entire spectrum of order-k $v^{(k)}$ to find a large enough k covering $P' \subset P$ and leaving $P - P'$ to be covered by the additional department. Figure 7(b) shows the $v\left(P^{(10)}\right)$. Here, $V^{(10)}$ of two police departments are green shaded and their ten nearest murder incidents are highlighted in thick black crosses. Note that, some murder incidents are covered by both departments' ten nearest crime incidents. In this case, both police departments will cooperate in the investigation. There are 11 incidents not covered by either department when k is 10. These uncovered incidents are scattered around the study region and we cannot easily spot a candidate location within the study region for the additional police department that will cover the other incidents.

Figure 7(c) shows the $v\left(P^{(15)}\right)$ where each police department investigates 15 murder incidents at the same time. In this case, seven murder cases are left uncovered with two cases in the north, two in the east, and three in the south. An order-15 Voronoi region shaded in yellow in the eastern part of Fig. 7(c) covers the seven remaining incidents. However, this region falls outside of the study region. Figure 7(d) shows the $v\left(P^{(17)}\right)$ which does not improve the coverage of existing police departments. There are still seven incidents uncovered. However, there exists a $v\left(P^{(17)}\right)$ (shaded in yellow) which covers the seven remaining incidents and falls within the study region. This suggests that the minimum k would be 17 and the possible location of the additional department would be the intersection area of the study region and the yellow shaded $v\left(P^{(17)}\right)$ as shown in Fig. 7(d).

4.2 Case Study 2

Brisbane is the capital city of Queensland, and similar to other major cities, this study region is experiencing rapid urban development. The local government is considering relocation of one of the police departments without affecting the order of *k*murder investigations currently undertaken by each police department. Currently, each department investigates murder incidents in the following order: the nearest murder first, the second next and so on and so forth up to *k*. Our proposed system can strategically support this type of question.

This case study is particularly useful when locations of target interest (police stations in this example) are constantly changing. This task can be modeled by the ordered order-*k* Voronoi diagrams. Figure 8(a) shows $v\left(P^{<3>}\right)$ when *k* is 3. The shaded Voronoi regions represent areas where police stations can move around without affecting the order of investigations.

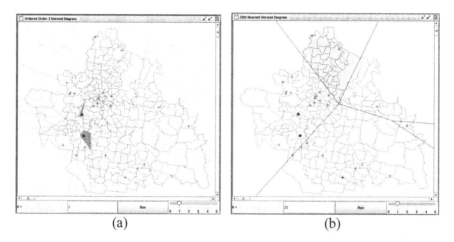

(a)   (b)

Fig. 8 Ordered order-*k* and *k*-th nearest Voronoi diagrams of *P*: (a) $v\left(P^{<3>}\right)$; (b) $v\left(P^{[25]}\right)$

4.3 Case Study 3

Murder is one of the worst crime types and families of murder victims may want to move to a new location to start a new life away from the constant reminders of the criminal act. In this case, they may want to move as far away as possible from their corresponding incident(s). In this case study, the local government is adopting a relocation program that repositions families of murder victims wanting to transfer to another place within the study region. The main aim of this relocation program is to displace the families within the study region having the corresponding murder case as the *k*-th nearest (farthest).

In this case study, we consider a family moving within the defined geographical boundaries. Note that, $v\left(P^{<k>}\right)$ becomes the farthest Voronoi diagram when $k = |P|$. Thus, this scenario can be modeled by the *k*-th nearest Voronoi diagram. Figure 8(b) shows $v\left(P^{[25]}\right)$. The shaded areas are suburbs where the family of the murdered individual could move to. Considering that the incident is at the southern part of the study region (marked +), they can relocate having the site of incident as the farthest point in the study region. This type of planning could speed up the time taken for individuals and families to integrate back into their communities.

5 Further Development into Minkowski Metrics

In our system, we have assumed the distance metric as the Euclidean distance, so-called crow-flies-distance. The Euclidean distance is an instance of the Minkowski metric and in urban geography another instance of the Minkowski metric, the Manhattan distance, better approximates real world situations [27]. We utilize three instances of the Minkowski metric, namely $p = 1$, $p = 2$, and $p = \infty$. Let $d_{L_p} : \Re^2 \times \Re^2 \to \Re$ denote the Minkowski metric defined by

$$d_{L_p}(r,s) = \left(\sum |r_i - s_i|^p \right)^{1/p} \ (p \in \Re \cup \{\infty\}, p \geq 1). \tag{4}$$

If $p = 1$, then $d_{L_1}(r,s) = |r_x - s_x| + |r_y - s_y|$ is the Manhattan metric, the taxicab metric, the city-block metric or the rectilinear metric [18]. The Minkowski metric

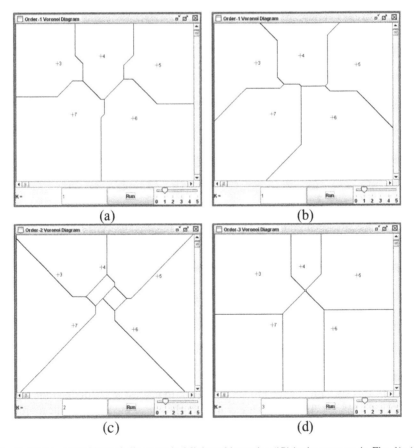

Fig. 9 Higher order Voronoi diagrams in Minkowski metrics ($|P|$ is the same as in Fig. 1): (**a**) $v_{L_1}^{(1)}(P)$; (**b**) $v_{L_\infty}^{(1)}(P)$; (**c**) $v_{L_\infty}^{(2)}(P)$; (**d**) $v_{L_1}^{(3)}(P)$

becomes the Euclidean metric when $p = 2$. If $p = \infty$, then the Minkowski metric becomes $d_{L_\infty}(r,s) = \max \left\{ |r_x - s_x|, |r_y - s_y| \right\}$ which is called the dominance metric, the maximum metric or the supremum metric [18]. We denote by v_{L_p} the Voronoi diagram in the Minkowski metric d_{L_p}.

We can extend our framework to Minkowski metrics to better model urban geography and to extend to other situations. In this case, enclosing shapes of Delaunay triangles are not circles, but squares for the Maximum metric and rhombuses for the Manhattan metric [18, 28]. Once we maintain order-k Delaunay triangles with these metrics, the same algorithm described in Section 3.3 can be applied to retrieve higher order Voronoi diagram in the Minkowski metrics. Figure 9 illustrates some of these diagrams. With this extension, it is possible to have what-if emergency response analysis with different Minkowski metrics.

6 Final Remarks

In this chapter, we introduce a robust framework for what-if emergency response management based on our previous work [29]. In this extended work, we explain how our proposed framework can be used to systematically support different types of what-if analysis. It enables the user to explore the complete order-k, ordered order-k and k-th nearest Voronoi diagrams that provide on-the-fly information for various what-if queries and what-if scenarios. We present several scenarios for how it can be used for various stages of emergency management. Undoubtedly, it can be used for the four phases (mitigation, preparedness, response, and recovery) of emergency management [30]. It needs to build a unified order-k Delaunay triangle data structure to start with. Once constructed, it can process various topological and region queries in diverse situations, and can support dynamic updates for interactive emergency analysis.

Our system can be extended to include more what-if scenarios in parallel with other Voronoi diagram generalizations. In some situations, professional bodies or disasters may have different weights (resources, impacts or damages). These can be modeled by weighted Voronoi diagrams. In some evacuation situations, the "crow-flies-distance" may not be the right choice since obstacles may block roads to evacuation places and simply roads are not straight. The constrained Delaunay triangulation seems to be a solid candidate for these scenarios. A further study is required to extend our data structure to handle all these scenarios.

References

1. Jayaraman V, Chandrasekhar M, Rao U: Managing the Natural Disasters from Space Technology Inputs. Acta Astronautica 40 (1997) 291–325
2. Alexander D: Principles of Emergency Planning and Management. Oxford University Press, New York (2002)
3. Canton LG: Emergency Management: Concepts and Strategies for Effective Programs. Wiley, New Jersey (2007)

4. Haddow G, Bullock J, Coppola DP: Emergency Management. 2nd edn. Elsevier, Oxford (2007)
5. Chang N, Wei YL, Tseng CC, Kao CYJ: The Design of a GIS-Based Decision Support System for Chemical Emergency Preparedness and Response in an Urban Environment. Computers, Environment and Urban Systems 21(1) (1997) 67–94.
6. Dymon UJ, Winter NL: Evacuation Mapping: The Utility of Guidelines. Disasters 17(1) (1993) 12–24
7. Goodchild MF: GIS and Disasters: Planning for Catastrophe. Computers, Environment and Urban Systems 30(3) (2006) 227–229
8. Kevany MJ: GIS in the World Trade Center Attack – Trial by Fire. Computers, Environment and Urban Systems 27(6) (2003) 571–583
9. Montoya L: Geo-Data Acquisition Through Mobile GIS and Digital Video: An Urban Disaster Management Perspective. Environmental Modelling & Software 18(10) (2003) 869–876
10. Salt CA, Dunsmore MC: Development of a Spatial Decision Support System for Post-Emergency Management of Radioactively Contaminated Land. Journal of Environmental Management 58(3) (2000) 169–178
11. Zerger A, Smith DI: Impediments to Using GIS for Real-Time Disaster Decision Support. Computers, Environment and Urban Systems 27(2) (2003) 123–141
12. Gahegan M, Lee I: Data Structures and Algorithms to Support Interactive Spatial Analysis Using Dynamic Voronoi Diagrams. Computers, Environments and Urban Systems 24(6) (2000) 509–537
13. Balasubramanian V, Massaguer D, Mehrotra S, Venkatasubramanian N: DrillSim: A Simulation Framework for Emergency Response Drills. In Mehrotra S, Zeng DD, Chen H, Thuraisingham BM, Wang FY, eds.: Proceedings of the IEEE International Conference on Intelligence and Security Informatics. Lecture Notes in Computer Science 3975, Springer, San Diego, CA (2006) 237–248
14. AEATechnology: EGRESS.   http://www.aeat-safety-and-   risk.com/html/egress.html (Accessed in 2006)
15. CrowdDynamics: Myriad II. http://www.crowddynamics.com (Accessed in 2006)
16. IES: Virtual Environment. http://www.iesve.com (Accessed in 2006)
17. Lee I, Gahegan M: Interactive Analysis Using Voronoi Diagrams: Algorithms to Support Dynamic Update from a Generic Triangle-Based Data Structure. Transactions in GIS 6(2) (2002) 89–114
18. Okabe A, Boots BN, Sugihara K, Chiu SN: Spatial Tessellations: Concepts and Applications of Voronoi Diagrams. 2nd edn. John Wiley & Sons, West Sussex (2000)
19. Gold CM: Problems with Handling Spatial Data – The Voronoi Approach. Canadian Institute of Surveying and Mapping Journal 45(1) (1991) 65–80
20. Aurenhammer F, Schwarzkopf O: A Simple On-Line Randomized Incremental Algorithm for Computing Higher Order Voronoi Diagrams. In: Proceedings of the Seventh Annual Symposium on Computational Geometry, ACM Press, North Conway, NH (1991) 142–151
21. Chazelle B, Edelsbrunner H: An Improved Algorithm for Constructing kth-Order Voronoi Diagrams. IEEE Transactions on Computers 36(11) (1987) 1349–1354
22. Dehne FKHA: An O(N4) Algorithm to Construct all Voronoi Diagrams for K-Nearest Neighbor Searching in the Euclidean Plane. In Diaz J, ed.: Proceedings of the International Colloquium on Automata, Languages and Programming. Lecture Notes in Computer Science 154, Springer, Barcelona, Spain (1983) 160–172
23. Lee DT: On k-Nearest Neighbor Voronoi Diagrams in the Plane. IEEE Transactions on Computers 31(6) (1982) 478–487
24. de Berg MT, van Kreveld MJ, Overmars MH, Schwarzkopf O: Computational Geometry: Algorithms and Applications. 2nd edn. Springer-Verlag, Heidelberg (2000)
25. Kraak MJ. Visualising Spatial Distributions. John Wiley & Sons Inc., New York (1998)
26. Murray AT, McGuffog I, Western JS, Mullins P: Exploratory Spatial Data Analysis Techniques for Examining Urban Crime. British Journal of Criminology 41(2001) 309–329

27. Krause EF: Taxicab Geometry. Addison-Wesley, California (1975)
28. Jünger M, Kaibel V, Thienel S: Computing Delaunay Triangulations in Manhattan and Maximum Metric. Technical Report 94.174, Institutfur Informatik, Universitat zu Koln (1994) [Available at http://math.tu-berlin.de/ kaibel/papers/mandel.ps.gz].
29. Lee I, Pershouse R, Phillips P, Christensen, C: What-if Emergency Management System: A Generalized Voronoi Diagram Approach. In: Proceedings of the Pacific Asia Workshop on Intelligence and Security Informatics. Lecture Notes in Computer Science 4430, Springer, Taipei, Taiwan (2007) 58–69
30. Haddow GD, Bullock JA: Introduction to Emergency Management. Butterworth-Heinemann, Stoneham, MA (2003)

Identity Management Architecture

Uwe Glässer and Mona Vajihollahi

*Everything is vague to a degree you do not realize
till you have tried to make it precise.*

Bertrand Russell, 1918

Abstract Identity management plays a crucial role in many application contexts, including e-government, e-commerce, business intelligence, investigation, and homeland security. The variety of approaches to and techniques for identity management, while addressing some of the challenges, has introduced new problems especially concerning interoperability and privacy. As such, any attempt to consolidate such diverse views and approaches to identity management in a systematic fashion requires a precise and rigorous unifying semantic framework. We propose here a firm semantic foundation for the systematic study of identity management and improved accuracy in reasoning about key properties in identity management system design. The proposed framework is built upon essential concepts of identity management and serves as a starting point for bringing together different approaches in a coherent and consistent manner.

Keywords Identity theft · Identity resolution · Information sharing · Privacy and trust · Semantic modeling

1 Introduction

Identity and *identity management* are two key concepts that have been addressed by researchers from different disciplines with different methods in various application domains. Across disciplines, there is a consensus on the vital role of identity management in many strategic applications, including investigation contexts, services provided by governments (e-government), e-commerce, business intelligence, and

Uwe Glässer and Mona Vajihollahi
Software Technology Lab, School of Computing Science, Simon Fraser University, Canada
e-mail: {glaesser,monav}@cs.sfu.ca

C.C. Yang et al. (eds.), *Security Informatics*, Annals of Information Systems 9,
DOI 10.1007/978-1-4419-1325-8_6, © Springer Science+Business Media, LLC 2010

homeland security, although the focus in each context is different. For instance, social scientists are mainly interested in the theoretical discussion on "what is identity?" and "what constitutes identity?" [21], while in the *digital world* context, the main focus is on "digital identity" and its required elements [3]. One of the most challenging issues related to identity management is *privacy*. Privacy-related requirements impose important restrictions on the design and effectiveness of any identity management system and hence are of utmost importance [5]. At a more technical level, there are a number of outstanding issues, including resolution techniques [17, 14, 25], centralized, distributed, or federated design of systems [29], and differences between *identification* and *anonymous authentication* [3].

Although there have been substantial efforts to address the challenges in each of the above-mentioned areas, the sad reality is that there is no common agreement or understanding on even the basic concepts, such as what constitutes identity or what is identification. In other words, in the absence of a common starting point for research on identity, different assumptions have been used by different researchers in solving specific problems. The lack of such a unifying view has several negative implications including, but not limited to,

- interoperability problems (especially within government or surveillance contexts);
- privacy-related issues (e.g., What should be protected? What is "personal data?" Interchange data format in health applications);
- issues related to reconstructing identities for investigation or profiling purposes (e.g., legitimate vs. illegitimate profiling);
- difficulty in bringing together research results (from different areas) on this topic.

In recent years, the need for a well-designed identity management system (IMS) has widely been recognized by different groups working on identity management around the globe, namely there have been several initiatives in Europe to address this issue. The study "Identity Management Systems (IMS): Identification and Comparison" [12] provides an overview of the open issues in the field. FIDIS (Future of Identity in Information Society) [9] and PRIME (Privacy and Identity Management for Europe) [19] are two research initiatives by the European Union toward advanced research on identity management. Challenges include establishing proper associations between entities (civilians or institutions) and their identities, matching of identities, and detecting fake identities. In the aftermath of September 11, 2001, even more attention has been directed to this area to provide governments and intelligence agencies with better intelligence and better tools for identifying and responding to possible threats.

Building on our experience with semantic modeling of behavioral aspects of complex distributed systems, such as semantic foundations of Web service architectures [8], in this project,[1] we focus on developing a firm unifying semantic

[1] This research project is funded by the Ministry of Labour & Citizens' Services of British Columbia, Canada.

foundation for a systematic study of identity management and improved accuracy in reasoning about key properties in IMS design. Given the diversity and variety of the concepts and disciplines involved, we argue that mathematical rigor and precision are essential in consolidating existing approaches and harmonizing the sharing and integration of information.

We propose a novel conceptual model of identity along with a simple, but universally applicable, mathematical framework for establishing a precise semantic foundation for the most fundamental elements of identity management. This model serves as a starting point for bringing together different approaches to identity management in a more systematic fashion. Through an extensive review of the literature, we also identify essential requirements of any IMS (such as privacy, user-control, and minimality) and focus on the practical relevance of developing a distributed approach to identity management (as opposed to a centralized one). Finally, we illustrate the practicality of our approach by applying the model to a rigorous definition of identity theft.

Section 2 presents an extensive review of the identity management literature, summarizing common identity management concepts and terminology. Section 3 introduces the formal semantic framework and Section 4 mainly focuses on the application of the model in identity theft. Section 5 concludes the chapter.

2 Background

Identity, identity management, and related issues are studied in different application contexts and across disciplines. Starting from the existing literature we explore here the basic definitions, common terminology, and also some of the related work on technical issues, such as identity resolution. We also provide an extensive review of existing work on identity-related issues, including privacy and identity theft.

2.1 Basic Definitions

As noted by Camp [3], "the word 'identity' refers to an increasing range of possible nouns . . . from a set of personality-defining traits to a hashed computer password." A glossary of the terminology used in digital identity systems is provided in [3], which is the result of a cooperative attempt to develop a common understanding of the terminology across several disciplines [4]. This glossary was developed to address the problem of the overload of identity terminology and "the lack of conceptual clarity" [3]. Hence, we consider it as one reference point in designing our model of identity.

In [11], Harper provides a slightly different taxonomy of the elements involved in an identity system. *Identifiers* are introduced as the building blocks of identification, and special attention is dedicated to the classification of identifiers into the following groups: *something-you-are*, *something-you-are-assigned*, *something-you-know*, and *something-you-have*.

Most of the works on digital identity, identity management, and identity matching, however, use a wide variety of terms not necessarily defined by any of the above-mentioned sources. For instance, in [6] the authors introduce the notion of "partial identities" which are subsets of a person's identity and uniquely identify the person. They also used the word "pseudonym" to refer to identifiers of subjects. Although these concepts proved to be useful in modeling an identity management system, they are not introduced or addressed in either [11] or [3].

For a complete list of definitions from these different sources, we refer to the glossary provided in our technical report [10].

2.2 Identity (Entity) Resolution

The problem of *matching* and *relating* identities fits under the broader Entity Resolution problem, which is a well-known issue in databases and data mining. In [14], a matching technique specific to identities, known as identity resolution, is described. This approach was originally proposed to address identity matching problems in Las Vegas casinos and is "designed to assemble i identity records from j data sources into k constructed, persistent identities." It uses a deterministic technique based on expert rules in combination with a probabilistic component to determine *generic* values for identifiers. Since generic values, such as the phone number of a travel agency, are widely used by different people, they are not relied upon in identity matching. The approach also detects relationships between identities and produces useful alerts about potentially dangerous ones. Identity resolution is sold as an off-the-shelf product by IBM and has been used in several application domains including gaming, retail, national security, and disaster response.

The Artificial Intelligence Lab at the University of Arizona focuses on developing algorithms that automatically detect false identities to assist police and intelligence investigations. Wang et al. [25] propose a record comparison method to address the problem of identifying deceptive criminal identities. The algorithm builds on a taxonomy of identity deception developed from a case study of real criminal deception patterns [26]. In [26], criminal deception patterns are categorized into four groups: name deception, residency deception, DPB (date/place of birth) deception, and ID deception. Focusing on these fields, the overall disagreement between two records is defined as the summation of the disagreements between their constructing fields. A supervised training process is used to determine an appropriate disagreement threshold for matching. Hence, there is a need for training data. In [27], a probabilistic Naïve Bayes model is proposed to address the same problem. The model uses the same four features for matching identities; however, a semi-supervised learning method is used that reduces the effort required for labeling training data.

Phiri and Agbinya propose a system for management of digital identities using techniques of information fusion [17]. Instead of relying on a single credential (e.g., PIN number or password) for authentication, they suggest a *multimode credential authentication* involving a combination of a number of credentials. In this approach, a set of credentials are presented by the user and the result of the information fusion process determines the outcome of authentication. However, training data is required to fine-tune the underlying neural network which performs the information fusion process.

2.3 Identity Management Systems

The issue of managing identities has become more important in recent years with the growth of the Internet and its wide variety of applications. Several attempts have been made to characterize the requirements of such identity management systems (IMS) focusing on the needs of both users and managers of the identity.

In [12], a comprehensive list of existing products and prototypes that provide user-controlled management of identities is studied. The study is built on four pillars: (1) basic IMS requirements, (2) usage scenarios, (3) evaluating identity management applications, and (4) survey of experts' expectations. IMS requirements are analyzed in three different contexts: social, legal, and technical, as shown in Fig. 1. For the purpose of usage scenarios, different application domains such as e-government, e-commerce, and e-health are considered. The study, however, concludes "none of the evaluated products meets all elaborated criteria," most notably "are significant deficiencies regarding privacy, security and liability." Consequently, experts have consistently scored privacy protection, security, and usability as the most essential features of IMS.

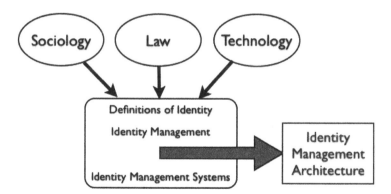

Fig. 1 Major contexts in defining identity and identity management [12]. Our proposed identity management architecture builds on basic definitions of identity and concepts of identity management to provide a firm semantic foundation for identity management systems.

The authors of [6] present an approach of developing an IMS with respect to multilateral security. They emphasize the role of IMS in realizing privacy laws and its importance for establishing user confidence in e-commerce applications. They also highlight the lack of *universal or standardized approaches* for IMS. The proposed system is built on top of SSONET, an architecture for multilateral security. Different usage scenarios are analyzed to define the requirements of a desired identity manager (including creation and management of pseudonyms, certification and managing credentials, user-control, and privacy). However, the issues regarding complexity and confidentiality of the identity manager remain open. Furthermore, the proposed identity manager relies on the secure functionalities provided by the SSONET framework.

In [7], identity management is regarded as "definitions and lifecycle management for digital identities and profiles, as well as environments for exchanging and validating such information." The paper presents the results of the Roadmap for Advanced Research in Privacy and Identity Management (RAPID) project. Similar to [6], the notion of *nyms* and partial identities are used in the project. The paper highlights three basic requirements of an IMS: (1) reliability and dependability, (2) controlled information disclosure, and (3) mobility support. The obstacles and open problems in developing a system that supports multiple and dependable digital identity (MDDI) are also explored.

In [22], the issues of rapid changes in the behavior of Internet users and the consequent challenges of dealing with new (and sometimes conflicting) requirements are discussed. The paper analyzes some new trends and the respective needs of the end users. One of the important needs of end users is having "multiple identities that are also streamlined and portable." Portability of the identities is a valuable feature, especially for users of reputation-based services such as eBay or Craig's list. Another issue is preserving privacy while highly personalized information is revealed in many networking Web sites. There is also the problem of *dispatching* or *killing* an identity, which is an important part of the digital identity lifecycle. The authors also point to the interesting phenomenon of *generational shift* and how the behavior and needs of the younger generation are different from the previous generation. This shift introduces new challenges for design of identity management systems, as users are no longer cautious or concerned about revealing their personal information.

2.4 Identity Theft and Identity Fraud

Broadly speaking, identity theft or identity fraud is defined as the misuse of another person's identity information. With a rapid growth in recent years, it has attracted much attention by the law enforcement agencies in the United States and Canada. According to a report by a bi-national working group to the Minister of Public Safety and Emergency Preparedness, Canada, and the Attorney General of the United States [20], "during a one-year period in 2002–2003, total losses to individuals and businesses related to identity theft in the United States were estimated at approximately US$53 billion. In Canada the losses for 2002 were estimated at approximately CAN$2.5 billion."[2] Despite all the attention and public concern,

---

[2] According to a 2007 report, identity theft loss in the United States declined to $49.3 billion in 2006, due to an increased vigilance among consumers and businesses [13].

there is little agreement among researchers, practitioners, and law enforcement officials on a clear definition of identity theft and identity fraud.

Sproule and Archer address this problem by developing a conceptual process model of the problem domain in a systematic way by consulting different stakeholders [23]. The process model classifies activities related to identity theft and fraud into different groups, which contribute to each phase of the process. The process starts with collection of personal information or creation of a fictitious identity/manipulation of one's own identity. The collected information is then used to develop a false identity or is directly used in committing crimes enabled by a false identity.

In [28], a contextual framework for identity theft is presented. The framework identifies the major stakeholders in the identity theft domain, the interactions between them and how the information flows. The stakeholders include (1) identity owners, (2) identity issuers, (3) identity checkers, (4) identity protectors, and (5) identity thieves. The paper also classifies the activities important in combating identity theft and identifies open problems of the field. Although the framework facilitates harmonizing and integrating research, and development in the identity theft problem domain, it does not provide a precise definition of the fundamental concepts such as identity and identity theft. For instance, the term "identity" is loosely used in some cases to refer to "identity information" (or personal information).

2.5 Privacy and Trust

The issues of safety, privacy, and trust in the digital world, and more particularly on the Internet, are strongly linked to identity problems and the lack of a native identity layer [2].

> The existing identity infrastructure of the Internet is no longer sustainable. The level of fraudulent activity online has grown exponentially over the years and is now threatening to cripple e-commerce. Something must be done now before consumer confidence and trust in online activities are so diminished to lead to its demise [5].

One proposed solution is designing a unifying identity metasystem that provides interoperability among underlying identity systems [2]. To capture the pragmatic requirements of such a system, a set of principles, called "Laws of Identity" are developed through an open consensus process among experts and stakeholders [2]. It is argued in [5] that such an identity metasystem significantly contributes to improving security and privacy in the online world.

In general the issue of privacy is closely related to identification. Hence, any attempt to integrate privacy protection into a system must address identity-related issues. This is addressed by a multi-national project, called PISA (Privacy Incorporated Software Agent) [18]. The project aims at identifying possible threats to the privacy of individuals resulting from the use of agent technologies and demonstrating ways of applying privacy-enhancing technologies (PET) to agent technologies in order to eliminate the impact of these threats. Even the most basic definitions such as personal data, identifiable data subjects, and identification are essential in defining privacy and identifying privacy threats. These issues and other results of the PISA project are addressed in [24].

2.6 Advanced Research on Identity Management

The Liberty Alliance project [16] aims at establishing open standards, guidelines, and best practices for identity management, mainly focusing on identity-based Web services. The project promotes a *federated identity management* approach which focuses on building *trust relationships* between business and the ability to *federate* isolated accounts of users among well-defined *trust circles* [15]. The goal is to develop open standards that address interoperability, management of privacy, and identity theft prevention.

Within the European Union there have been several multidisciplinary initiatives to studying digital identities, their management, and the related problems. The FIDIS (Future of Identity in Information Society) project [9] aims at integrating research efforts across different European nations focusing on challenging problems such as interoperability of identity and identification concepts, ID-theft, privacy, security, profiling, and forensic implications. One of the main research activities of the project focuses on exploring the definition of identity and identification. This research branch, called *Identity of Identity*, aims at developing an inventory of definitions in the identity domain and their respective use cases, analyzing existing models of identity management, and providing an overview of future directions of the models. FIDIS defines seven research branches each focusing on one important aspect of identity, such as privacy, interoperability, and mobility.

The PRIME (Privacy and Identity Management for Europe) project [19] addresses the lack of an identity infrastructure for the Internet, identifies essential requirements such as security and privacy, and aims at defining the right balance of such requirements in emerging identity management architectures. Similar to [2] and [5], PRIME takes a *user-centric* approach to identity management. A high-level architecture for identity management is proposed and is accompanied by PRIME toolbox and middleware to support and implement user-centric privacy and identity management functionalities.

3 The Formal Model

This section presents a precise semantic model of identity management concepts. Based on the existing literature, we identify key concepts in the domain of identity management and formalize their intuitive meaning in abstract functional and operational terms. We start with the most basic definitions.

3.1 Basics

In principle, the term *entity* as used in the following may refer to either an individual or an organization existing in the real world. In this chapter, however, the focus is on individuals rather than organizations.

> *Attribute.* A characteristic associated with an entity such as name, date of birth, height, fingerprints, iris scan, and genetic code. [3. 4].
>
> *Identity.* We define identity as an abstract (mental) picture of an entity, such that, for the identity management concepts considered here, an entity's identity is logically equivalent to the physical presence of this particular entity. In this view, any distinction between an entity and its associated identity is irrelevant and will not further be distinguished.
>
> *Partial identity.* Any subset of properties (read attributes) associated with users (read entity), such as name, age, credit card number, or employee number, that the user (entity) uses for interacting with other parties [7].[3]
>
> *Identifier.* An identifier identifies a distinct person, place, or thing within the context of a specific namespace [3, 4]. There are two types of identifiers[4]:
>
> - *Personal identifier:* A persistent identifier associated with an individual human consisting of one or more attributes that are difficult or impossible to change, such as date of birth, height, and genetic code.
> - *Pseudonym:* An identifier associated with non-persistent attributes or sets of transactions with no permanent or personal identifier.
>
> *Context.* A context refers to a specific application domain or circumstance in which a partial identity is defined and has a meaning.[5] Henceforth, we associate with each partial identity the specific context in which this partial identity is defined.

In the formal model, we regard an *identity* as the abstract representation of the corresponding *entity*. All the *attributes* of the entity then help in defining its identity. Therefore, the main building blocks of our model are as follows.

---

[3] In reality, one commonly uses combinations of characteristics in order to distinguish an entity from other entities, so that it becomes identifiable based on a certain set of attributes; however, it seems virtually impossible to find any such set that is generally suitable as a placeholder for an entity's identity in an absolute sense as assumed here.

[4] For the purpose of the first abstract model we do not distinguish between personal identifiers and pseudonyms.

[5] Several contexts may come together under the umbrella of a *domain*. For instance, several contexts exist within the health domain, including hospital records, health care providers, etc.

universe IDENTITY   // abstract representation of all individuals
universe ATTRIBUTE
universe CONTEXT
universe PARTIAL_IDENTITY
universe IDENTIFIER ≡ $\mathcal{P}(\text{ATTRIBUTE})$
// $\mathcal{P}$ denotes the power set, i.e. the set of all subsets of a given set.

attributeSet : IDENTITY → $\mathcal{P}(\text{ATTRIBUTE})$
// The set of all attributes that belong to an identity

A number of identifying functions exist that map an identifier to its respective partial identity within a context. These functions play the role of keys in databases.

g : IDENTIFIER × CONTEXT → PARTIAL_IDENTITY

A partial identity is used to *represent* an identity within a distinguished context. In other words, a partial identity is the "realization" of its respective identity within the context. Hence, the ideal is to always have a one-to-one relation between partial identities and real identities. In fact, an assumption often made by IMS designers is that a new partial identity is created only if the respective identity is not already represented in the context by another partial identity. However, this assumption is frequently violated in real-world applications, which is the cause of many of the core problems in identity management.

It is important to emphasize that from a context point of view the "real" (actual) association between partial identities and identities typically remains hidden; that is, there is no feasible way of checking or validating that the association is a correct one in an absolute sense. For our modeling purposes, however, we assume to have this absolute knowledge, although this possibility serves for illustrative purposes only. We therefore introduce an *oracle* that provides that hidden information. The oracle is defined as the relation[6] O and is meant to provide the real identity(ies) behind a partial identity. In each state of the system, the oracle maps a partial identity defined within a context to the identity that is represented by that partial identity.

$O \subseteq$ CONTEXT × PARTIAL_IDENTITY × IDENTITY

As noted above, in any given context, the ideal situation is to have a one-to-one mapping between partial identities and real identities. We thus define the following integrity constraints for the oracle. Of course, these constraints refer to the ideal situation and may (and often do) not hold in reality.

$$O \text{ is meant to be a mathematical function.} \tag{1}$$

That is within one context any single partial identity cannot represent more than one individual identity.

$$O \text{ is meant to be injective.} \tag{2}$$

---

[6] It is important to note that the oracle is not necessarily a function.

The second constraint means that two partial identities existing in the same context cannot be mapped to the same identity.

3.2 Mapping Partial Identities to Identities

Figure 2 illustrates four basically different cases of mapping partial identities to identities, some of which are potentially problematic. Note that in any given state, the oracle represents these mappings. In other words, for each context, the oracle has an understanding similar to the one shown in the diagram. Here, we describe each case in more detail.

Fig. 2 Mapping partial identities to identities within the same context.

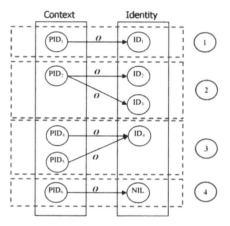

1. *Valid mapping*: In this case, a given partial identity PID_1 is uniquely mapped to an existing identity ID_1. In other words, the oracle knows that ID_1 is behind the partial identity PID_1.
2. *O is not a (mathematical) function*: PID_2 represents both ID_2 and ID_3. This happens when sufficient information about entities does not exist in a context. Hence, one partial identity is created which maps to two actual identities. This is obviously NOT a desired situation and violates the integrity constraint 1.
3. *O is not injective*: In this case, one identity is represented by several partial identities. This violates the injectivity constraint 2.
4. *Fake (fictitious) identity*: This case captures situations where there is no real identity behind a partial identity.

Cases 2–4 capture different *undesired* situations with respect to partial identities within one context. These cases can be categorized as *logical* inconsistencies and are explained using the mathematical representation. When analyzing partial identities across contexts, one can identify a second category of inconsistencies called *semantic* inconsistencies. In this case, within one context a mathematically valid

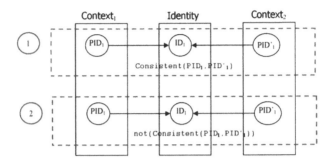

Fig. 3 Mapping partial identities to identities across contexts

mapping exists between a partial identity and its respective identities (i.e., Case 1 in Fig. 2 applies); however, an inconsistency can be detected when other contexts are included in the analysis. Figure 3 shows two different cases of mapping of two partial identities across different contexts. The first case shows a valid mapping across contexts: PID_1 and $PID\bullet_1$ are mapped to the same identity ID_1 and thus need to be semantically consistent; i.e., there is no inconsistency in their constituting attributes. In the second case, however, there is inconsistency between PID_1 and $PID\bullet_1$ in the sense that respective attributes have different values. Semantic inconsistencies within one context, such as having outdated records in one context, contribute to this type of inconsistencies across contexts.

One of the goals in identity management is to identify these cases, deal with them, and minimize their occurrence. This is further discussed under evolution of partial identities in the next section.

3.3 Evolution of Partial Identities

In general, the set of partial identities defined within a context changes dynamically over time. New partial identities are introduced, existing ones are discarded (killed [22]) or merged to form new partial identities, or one partial identity is split into two. These changes define the *life cycle* of partial identities within a context.

3.3.1 Identity Resolution

We consider identity resolution as the *internal* process of changing the state of a context with respect to its partial identities. These changes are meant to improve the overall *quality* of the mapping between partial identities and identities and, as a result, target the problematic cases discussed in Section 3.2. Several techniques have been proposed to identify and resolve these problems [17, 14, 25]. We provide a high-level view of the identity resolution concept, abstracting from the operational details of how and when it is performed and focusing on the semantic aspects.

Formally speaking, identity resolution is required when a violation of one of the integrity constraints (1) or (2) is detected. From the perspective of an outside observer, the state of a context changes after performing an identity resolution operation. Here we split the identity resolution process into two sub-tasks, each dealing with one type of integrity constraint.

$$\text{IdentityResolution} \equiv \text{FunctionResolve} \wedge \text{InjectvtyResolve}$$

At the highest level of abstraction, we describe the effect of each operation in terms of pre-conditions and post-conditions that respectively hold before and after performing each task.[7]

The function resolution operation is triggered when a new piece of information is obtained that reveals that one single partial identity is referring to two or more identities.[8] In our model, this new information is abstractly represented by the function *newInfo*. Assuming *idSet* $(c, p_1, newInfo)$ refers to the set of identities that, based on *newInfo*, are all represented by a single partial identity p_1 in the context c, the FunctionResolve operation is specified as follows.

pre $\exists id_1, id_2 \in idSet(c, p_1, newInfo)$
$\quad O(c, p_1, id_1) \wedge O(c, p_1, id_2) \wedge id_1 \neq id_2$

FunctionResolve(c : CONObjECT, p_1 : PARTIAL_IDENTITY, $newInfo$: INFO)

post $\forall id_1, id_2 \in idSet(c, p_1, newInfo), \forall p \in$ PARTIAL_IDENTITY
$\quad O(c, p, id_1) \wedge O(c, p, id_2) \Leftrightarrow id_1 = id_2$

The injectivity resolution operation is triggered when a set of partial identities *pSet* are detected, all of which refer to the same identity id_1. Therefore, the InjectvtyResolve operation is specified as follows.

pre $\exists\, id_1 \in$ IDENTITY, $\forall p \in pSet \quad O(c, p, id_1)$

InjectvtyResolve(c : CONTEXT, $pSet$: $\mathcal{P}$(PARTIAL_IDENTITY))

post $\forall p_1, p_2 \in$ PARTIAL_IDENTITY
$\quad O(c, p_1, id_1) \wedge O(c, p_2, id_1) \Leftrightarrow p_1 = p_2$

It is worth mentioning that the existing literature does not provide a precise high-level definition of identity resolution. Instead, in a bottom-up fashion, identity resolution is commonly defined by describing the underlying algorithms that perform the operation. This is a potential source of inconsistencies, especially when interoperability is a priority. To address this problem we illustrate how one can properly specify each of the above-mentioned operations in a

---

[7] We define here a more general case with n identities/partial identities.

[8] In practice, different heuristic approaches and AI-based techniques are used to extract this information [17, 13, 25].

systematic fashion, such that they can be further refined into appropriate algo-rithms and implementations. While being simple and easy to understand, the formalization used here is precise and grounded in mathematical logic based on the *abstract state machine* (ASM) formalism and abstraction principles [1].[9]

FunctionResolve(c : CONTEXT, p_1 : PARTIAL_IDENTITY, $newInfo$: INFO) $\equiv$
   // Resolved by splitting p_1 into different partial ids
   Delete(c, p_1)
   SplitIntoNewPartialIDs($c, p_1, newInfo$)

The delete, split, ans merge operations are purposefully left abstract at this level. Details can be added as needed in systematic refinement steps.

InjectvtyResolve(c : CONTEXT, $pSet$: $\mathcal{P}$(PARTIAL_IDENTITY)) $\equiv$
   // Resolved by merging partial identities into one
   forall pid **in** $pSet$ **do** [*in parallel*] Delete(c, pid)
   MergeIntoANewPartialID($c, pSet$)

As a result of the resolution process, one or more partial identities are deleted from the context, and one or more new partial identities are created using the information from the old ones. Here, we introduce a new concept: the new partial identities may come with two additional attributes, respectively, specifying the *confidence* in the newly created partial identity and the *history* of its evolution (allowing to undo the operation if necessary). The confidence value is a normalized numerical value expressed as a real number in the interval [0,1].

4 Applications

In this section, we explore potential applications for which the semantic frame-work of the identity management architecture could be used and illustrate how it would be helpful. An important usage of the framework is for achieving inter-operability across different domains, which is impossible without a coherent and consistent semantic foundation. Here we can make an analogy to the programming languages domain and the importance of having a well-defined semantic for a given programming language. Otherwise, different implementations of the same language may result in different interpretations of some concepts of the language, leading to inconsistent behavior of the same program in different implementations. We also apply the framework in other contexts such as identity theft, privacy, and criminal investigations.

---

[9]For a comprehensive list of references on ASM theory and applications, we refer the reader to the ASM Research Center at www.asmcenter.org.

4.1 Identity Theft

Section 2.4 provides a brief overview of the literature on identity theft. The authors of [23] developed a conceptual model of identity theft and identity fraud by identifying related common activities, as shown in Fig. 4.

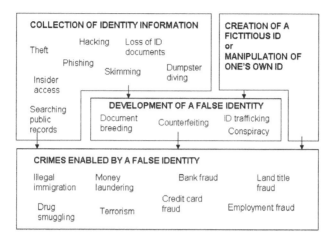

Fig. 4 Conceptual process model of identity theft domain from [23][10]

As a result, the following two definitions are provided [23]:

Identity theft: the unauthorized collection, possession, transfer, replication, or other manipulation of another person's personal information for the purpose of committing fraud or other crimes that involve the use of a false identity.

Identity fraud: the gaining of money, goods, services, other benefits, or the avoidance of obligations, through the use of a false identity.

A conceptual model clearly delineating identity theft and identity fraud is a valuable contribution and a good starting point for integrating research and practices in the field. However, we believe that such fundamental concepts need to be more rigorously defined. For instance, both definitions use the term *false identity*, which is only loosely defined and lacks a precise semantics. In fact, the fundamental question of "what constitutes an identity" is mostly neglected in the literature on identity theft. Consequently, it is difficult to infer what exactly a false identity is. Here, we use our core model of Section 3, extending this model to capture the notion of identity theft.

Our first goal in extending the formal framework is to better understand what exactly is considered a false identity. We derive our definition based on the process defined in [23] and the framework described in [28]. The latter defines five roles involved in identity theft: identity owner, identity checkers, identity issuers, identity protectors, and identity thieves (see Section 2.4).

---

[10] Courtesy of S. Sproule and N. Archer.

In our model of identity theft we introduce a new, more abstract role, called the Presenter of Identity Information. In principle, this entity is either the owner of the presented information or a third party. If the third party is *not authorized* to use the presented information, he/she is considered a thief. Hence, the *presented information* is a key element in defining a presenter. In Section 3, we explained how identities can be seen as abstract representations of entities. Here, we use this abstraction and define a presenter p as a tuple consisting of an identity id_p (the real person) and a set of attributes being presented $presAtrbts_p$ (the presented information).

$$\forall\, p \in \text{PRESENTER}\ \ p \equiv (id_p, presAtrbts_p)$$
$$\text{where } id_p \in \text{IDENTITY}\ \wedge\ presAtrbts_p \in \mathcal{P}(\text{ATTRIBUTE})$$

We also formalize the notion of *identification* using the definition provided in [3]. In the process of identification, a physical entity that presents a set of attributes[11] is *matched* to an *existing* identity.[12]

$$identification : \text{PRESENTER} \rightarrow \text{IDENTITY}$$

We can now precisely define the difference between a presenter's identity and the identification of a presenter, which is a source of ambiguity. The identity id_p is an integral part of the presenter; it is defined statically and cannot be detached from the presenter. Referring back to the model of Section 3, the $attributeSet(id_p)$ function provides the set of attributes associated with id_p. This set may change over time as the attributes of an identity may change, whereas the identity itself does not change. Identification of a presenter, on the other hand, is a dynamic process which is based on the *presented attributes* of the presenter ($presAtrbts_p$) not necessarily its *real attributes*– the ones that legitimately belong to the presenter ($attributeSet(id_p)$).

We abstractly define the process in which the presented attributes are mapped to an identity. A presenter p presents a set of attributes $presAtrbts_p$ to an *issuer*(or *checker*) of partial identities. This is where identification happens. As discussed in Section 3, within a given context *partial identities* are used as representations of identities. As such, the identification process logically splits into two consecutive steps: (1) mapping the attribute set to a partial identity in a given context and (2) implicit association between the partial identity and its respective identity. The following schema summarizes the identification process.

$$p \overset{presents}{\longrightarrow} presAtrbts_p \overset{matches}{\longrightarrow} pId \overset{represents}{\longrightarrow} id \tag{3}$$

We now explicitly define what is referred to as false identity in the literature. A false identity is assumed by a presenter when presenting attributes for identification that do not *belong* to the presenter's identity, as stated below.

---

[11] We are not concerned with the authentication of the attribute set and assume the attributes are authenticated.

[12] Note that if matching results in several identities, a logical inconsistency exists (see Case 2 in Fig. 2), which has to be resolved separately. Hence, we restrict here to one identity only.

$$falseIdentity(p) \Leftrightarrow presAtrbts_p \notin \mathcal{P}(attributeSet(id_p)) \tag{4}$$

As shown in Fig. 4, the conceptual process model of [23] identifies two different categories of activities that happen before a false identity is developed: (1) collection of personal information and (2) creation of fictitious identity or manipulation of one's own identity. However, only the activities in the first category are considered as identity theft. We want to address this issue in further detail using our formal model. In the following, we use the terms *id*, *pId*, and *matches* from schema 3.

In the first case, in a given context c, the collected information $presAtrbts_p$ forms a *false identity that matches a real identity id*; i.e.,

$$\exists \, pId \in \text{PARTIAL\_IDENTITY}, id \in \text{IDENTITY}$$
$$matches(c, presAtrbts_p) = pId \; \wedge \; O(c, pId, id)$$

It is important to point out that, in this case, it is *implicitly* assumed that the collection of personal information is not *authorized* by the owner of the identity information. We later discuss the implications of this assumption and the importance of making it explicit.

A *fictitious identity* is a false identity which is not based on a real person [23]. This happens when the result of the identification process yields $id = undef$; more precisely, for a presenter p in a given context c

$$\exists \, pId \in \text{PARTIAL\_IDENTITY}$$
$$matches(c, presAtrbts_p) = pId \; \wedge \; O(c, pId, undef) \; // \; O \text{ is oracle}$$

Therefore, it is safe to assume that identity theft does not occur in this case. However, if a false identity is created by manipulation of one's own identity, it may still match a real identity attributes (similar to the first case). Hence, we argue that identity theft can technically occur in this case.

In the following, we use our formal model to highlight some of the loose ends in the existing definition, arguing that there is a need for more precision and rigor in defining identity theft. In fact, we look at the problem from a different perspective and reorganize different cases as follows:

1. The presented information does not belong to the presenter and is mapped to another person's identity; i.e., $id_p \bullet id \bullet undef$

   – The presenter id_p has proper authorization from the owner of the identity information id; i.e., $isAuthorized(id_p, id, now)$ holds. It is important to point out the need for real-time evaluation of the authorization, which emphasizes the dynamic nature of the process.[13]
   – The presenter id_p does NOT have proper authorization from the owner id; i.e., *identity theft* occurs.

---

[13]Other factors, such as the specific context where identification occurs, should also be considered in authorization. However, for simplicity we use this broader definition of authorization.

2. The presented information leads to a fictitious identity; i.e., $id = undef$. Hence, identity theft does not occur.
3. Let us assume that a person, by mistake, presents his/her own attributes incorrectly. Since these attributes are not owned by that person, according to the definition (see Def. 4), a false identity is assumed.

   – Such a false identity may be fictitious, as described above (i.e., $id = undef$). In that case, identity theft does not occur.
   – If the false identity still maps to the right person (i.e., $id = id_p$), for instance due to the flexibility of the mapping algorithm, identity theft does not take place either.
   – However, if the false identity is NOT fictitious and does NOT map to the right person (i.e., $id_p \bullet id \bullet undef$), *identity theft* has taken place, by definition.

It is interesting to note the different insight that the formal definition provides. First, the notion of authorization introduces further complexity of real-time evaluation, context of authorization, and validation that deserve clarification in the definition of identity theft. Second, as far as identity theft is concerned, it is not important whether the presented information is a manipulated/modified version of one's own attributes or is completely stolen. If the presented attributes are mapped to an identity which is different from the identity of the presenter, an unauthorized use of the personal information of the person behind that identity has happened; hence, there is a case of identity theft. This important observation has not been clearly addressed before.

In the conceptual model of [23], it is implicitly assumed that "manipulation of one's own identity" is not malicious and is mostly used for preserving privacy. As a result, it is not included in the activities that contribute to identity theft. Part of the problem may have been caused by the vague definition of *one's identity*, which allows for using terms like "multiple identities" for one person, or "manipulation of one's own identity" without differentiating between identity and the attributes presented for identification.

Having a unified semantic framework facilitates integration across contexts and allows for *distributed* and *dynamic* analysis approaches for fraud detection. Within such a framework, one can identify and properly define different cases of identity theft and develop proper safeguards against misuse of identity.

4.2 Other Applications

In an investigation context, the identity management problem is viewed from a different perspective. The police, or the investigators, collect bits and pieces of information in order to derive the identity of a criminal offender. In other words, using the terminology of our framework, they try to collect enough *attributes* to reconstruct a *partial identity* which can be mapped to an *identity*. This is a highly dynamic process, since every new piece of information may change the constructed partial identity(ies) and the respective mappings. Crime investigation engines must deploy inference methods that oversee causal relationships between events and partial identities, time and location of crime events, and crime signatures. The high dynamics calls for systematic approaches to design of computer-assisted investigation tools. As such, we contend that our semantic framework can be utilized as the first building block in that direction.

Another area of importance is privacy-related requirements in identity management. Having an abstract architectural view facilitates analyzing the impact of using different privacy preserving techniques, such as anonymization, especially when multiple parties (contexts) are involved. We plan to further explore these topics in our future work.

5 Conclusions and Future Work

Identity management is a challenging problem in many strategic application domains, especially in security-related and Web-based contexts. Although a wide variety of solutions have been offered to address some of the key issues of identity management, there is still no common agreement on the most fundamental concepts, such as what constitutes an identity. Addressing the lack of such a unifying view we propose here a precise semantic model based on common notions and structures of computational logic and discrete mathematics in order to provide a well-defined framework, called *identity management architecture*, for analyzing and reasoning about identity management concepts and requirements on IMS. To exemplify the practicality of the framework in the systematic study of identity management, we apply the model to semantic aspects of identity theft, trying to clarify the underlying definition. In our future work, we plan to extend the framework to other critical application domains, such as investigation and privacy contexts. We also plan to apply the framework to study the semantics of some of the existing standards for identity management, e.g., the concept of federated identity management.

References

1. E. Börger and R. Stärk. Abstract State Machines: A Method for High-Level System Design and Analysis. Springer-Verlag, Berlin, 2003.
2. K. Cameron. The Laws of Identity [online], December 2005. Available: www.identityblog. com/?p=354.
3. J. L. Camp. Digital identity. Technology and Society Magazine, IEEE, 23:34–41, 2004.
4. L. J. Camp, et al. Identity in Digital Government: a research report of the Digital Government Civic Scenario Workshop, 2003. Research Report.
5. A. Cavoukian. 7 Laws of Identity: The Case for Privacy-Embedded Laws of Identity in the Digital Age, 2006.
6. S. Clauß and M. Köhntopp. Identity Management and its Support of Multilateral Security. Computer Networks, 37(2):205–219, 2001.
7. E. Damiani, S. D. C. di Vimercati, and P. Samarati. Managing multiple and dependable identities. Internet Computing, IEEE, 7:29–37, 2003.

8. R. Farahbod, U. Glässer, and M. Vajihollahi. An Abstract Machine Architecture for Web Service Based Business Process Management. International Journal of Business Process Integration and Management, 1:279–291, 2007.

9. Future of Identity in the Information Society – FIDIS (January 2008 – last visited) Website. [online]. Available: www.fidis.net .

10. U. Glässer and M. Vajihollahi. Identity Management Architecture. Technical Report SFU-CMPT-TR-2008-02, Simon Fraser University, February 2008.

11. J. Harper. Identity Crisis: How Identification Is Overused and Misunderstood. Cato Institute, 2006.

12. Independent Centre for Privacy Protection Schleswig-Holstein, Germany and Studio Genghini & Associati, Italy. Identity Management Systems (IMS): Identification and Comparison Study, 2003.

13. Javelin Strategy and Research, 2007 Identity Fraud Survey Report, February, 2007.

14. J. Jonas. Threat and fraud intelligence, Las Vegas style. Security & Privacy Magazine, IEEE, 4:28–34, 2006.

15. Liberty Alliance. Liberty Alliance Identity Federation Framework (ID-FF) 1.2 Specifications, December 2007.

16. Liberty Alliance Project (January 2008 – last visited) Website. [online]. Available: www.projectliberty.org

17. J. Phiri and J. Agbinya. Modelling and Information Fusion in Digital Identity Management Systems. In Networking, International Conference on Systems and International Conference on Mobile Communications and Learning Technologies, 2006. ICN/ICONS/MCL 2006, 181–187, 2006.

18. PISA – Privacy Incorporated Software Agent. Information Security, Privacy and Trust. (February 2008 – last visited) [online]. Available: www.iit-iti.nrc-cnrc.gc.ca/projectsprojets/pisa e.html.

19. PRIME – Privacy and Identity Management for Europe. (January 2008 – last visited) Website. [online]. Available: www.prime-project.eu .

20. Public Safety and Emergency Preparedness Canada. Report on Identity Theft, October 2004.

21. C. D. Raab. Perspectives on 'personal identity'. BT Technology Journal, 23:15–24, 2005.

22. C. Satchell, G. Shanks, S. Howard, and J. Murphy. Beyond security: Implications for the future of federated digital identity management systems. In OZCHI'06: Proceedings of the 20th Conference of the Computer–Human Interaction Special Interest Group (CHISIG) of Australia on Computer–Human Interaction: Design: Activities, Artefacts and Environments, ACM, New York, 313–316, 2006.

23. S. Sproule and N. Archer. Defining identity theft. In Eighth World Congress on the Management of eBusiness (WCMeB 2007), 20–31, 2007.

24. G. van Blarkom, J. Borking, J. Giezen, R. Coolen, and P. Verhaar. Handbook of Privacy and Privacy-Enhancing Technologies – The Case of Intelligent Software Agents. College bescherming persoonsgegevens, 2003.

25. G. Wang, H. Chen, J. Xu, and H. Atabakhsh. Automatically detecting criminal identity deception: an adaptive detection algorithm. IEEE Transactions on Systems, Man and Cybernetics, Part A, 36:988–999, 2006.

26. G. A. Wang, H. Atabakhsh, T. Petersen, and H. Chen. Discovering identity problems: A case study. In LNCS: Intelligence and Security Informatics. Springer, Berlin/Heidelberg, 2005.

27. G. A. Wang, H. Chen, and H. Atabakhsh. A probabilistic model for approximate identity matching. In J. A. B. Fortes and A. Macintosh, editors, Proceedings of the 7th Annual International Conference on Digital Government Research, DG.O 2006, San Diego, CA, May 21–24, 2006, 462–463. Digital Government Research Center, 2006.

28. W. Wang, Y. Yuan, and N. Archer. A contextual framework for combating identity theft. Security & Privacy Magazine, IEEE, 4:30–38, 2006.

29. P. J. Windley. Digital Identity, chapter Federating Identity. O'Reilly, Sebastopol, CA, 118–142, 2005.

A Game Theoretic Framework for Multi-agent Deployment in Intrusion Detection Systems

Yi-Ming Chen, Dachrahn Wu, and Cheng-Kuang Wu

Abstract Due to cost considerations there must exist in intrusion detection system, a trade-off between the user's ease of access and capability of detecting attacks. The proposed framework applies two game theoretic models for economic deployment of intrusion detection agent. The first scheme models and analyzes the interaction behaviors between an attacker and intrusion detection agent within a non-cooperative game, and then the security risk value is derived from the mixed strategy Nash equilibrium. The second scheme uses the security risk value to compute the Shapley value of intrusion detection agent while considering the various threat levels. Therefore, the efficient agent allocation creates a minimum set of deployment costs. The experimental results show that with the proposed two-stage game theoretic model, the network administrator can quantitatively evaluate the security risk of each IDS agent and easily select the most critical and effective IDS agent deployment to meet the various threat levels to the network.

Keywords Nash equilibrium · Shapley value · Security risk value · Agent deployment game

1 Introduction

Computer networks have become increasingly vulnerable and attacks on them have simultaneously grown in complexity and automation. Since network attacks can easily put e-commerce organizations out of business, intrusion detection systems must be devised to reduce network damage. To meet the threat of network attacks,

Yi-Ming Chen, Cheng-Kuang Wu
Department of Information Management, National Central University, Taiwan, ROC
e-mail: cym@mgt.ncu.edu.tw
Dachrahn Wu
Department of Economics, National Central University, Taiwan, ROC
e-mail: drwu@mgt.ncu.edu.tw

C.C. Yang et al. (eds.), *Security Informatics*, Annals of Information Systems 9,
DOI 10.1007/978-1-4419-1325-8_7, © Springer Science+Business Media, LLC 2010

enterprises and organizations have built security operation centers to control and analyze intrusion events. For example the Symantec ThreatCon warning system provides recommended protective measures, according to the threat level, that allows the system administrator to prevent, prepare for, mitigate, and respond to network attacks [22]. The several threat levels or conditions have a corresponding identification color, which indicates that mandatory protective measures have to be taken by the department administrators. However, these systems lack specific measures for rational decision-making, and do not use mathematical models to capture the interaction between attacker and defender. In this chapter, therefore, we propose a rating approach and an adjustable scale for preferences, which indicates the most suitable deployment of IDS given the threat level.

Recent intrusion detection systems can be divided into two classes: reactive preventing (e.g., signature or anomaly-based detection, ingress or egress filtering, etc.) and proactive mitigating (e.g., secure overlay services, proxy networks, etc.) [15]. Secure overlay services (SOS) architecture is for the prevention of network attacks and proactively mitigates distributed denial of service (DDoS) attacks, but secure agents require a large overhead for directing communication, and are limited in terms of its extending architecture [8]. Although SOS is more effective than reactive schemes, it still exists efficient in deployment problems. For example, how many overlay agents need to be deployed in the SOS architecture, and which agents are the best defenders for resisting attacks for the specific cost of IDS?

IDS architectures are prone to single points of failure that are vulnerable to discovery by attackers [8]. In order to solve this problem, as a back-up mechanism, it is necessary to designate many IDS agents so as to achieve the false tolerance in the secure overlay network. However, the more agents a security system deploys, the more vulnerability there is to discover and hence the price to pay is higher. Mass deployment and rich connectivity are not favorable for security. On the contrary, the fewer the agents deployed, the easier it is for attackers to penetrate, and the more difficult it is to respond. The robustness of a complete IDS decision process depends upon a balance between the agent requirements and failure tolerance. Therefore, there exists in IDS a trade-off between ease of user access and the capability of detecting IDS agents [1]. The administrator must develop a preference scale for improving agent deployment decisions and analysis so as to maintain equivalence between strategic decision-making and command-and-control in IDS management.

With the present optimization methods it is difficult to solve for IDS agent cost allocation problem (CAP) [23]. Hence, it is necessary to quantize the proposed framework for modeling attacks, to examine security risks (i.e., threats of attack), and to make decisions about actions in response. Here we use an economical solution for agent deployment. The game theoretic tools have been successfully applied in economics, political science, psychology, and decision theory. It also provides analytical techniques of use in many areas of research where multi-agents compete and interact with each other in a system [6]. In most multi-agent interaction, the overall outcomes depend on the choices made by all self-interested agents and make the choice that optimizes its outcome [18]. Therefore, we apply the game theoretic approaches to produce the framework for interaction between the network attacker and the IDS agent in relation to analysis and decisions related to deploying of the detection agent. It is a framework for deploying alarm multi-agent using game theory (DAMAG), which analyzes the detected threats and integrates with the existing IDS.

Interactions between the attacker and the IDS agent are modeled by game theory for identifying their behaviors in the intrusion detection systems. Additionally, the proposed DAMAG framework prioritizes to help the IDS agent make rational choices about severity of the condition, and to decide how many agents to deploy for the detection of significant threats in real time. Our study deploys an economical intrusion detection system which comprises two stages of game theoretic models for constructing two schemes. The first scheme models the interaction process between the network attacker and the IDS agent as a two-person and non-cooperative finite game. The proposed payoff functions utilize the security risk measures for two players (e.g., detection rate, false alarm rate, etc.). After this, the mixed strategy Nash equilibrium (NE) is derived from these functions and assigned as a unique security risk value (SRV) for the agent. Then, the second scheme constructed a cooperative IDS agent game. The power index is the Shapley value [17] applied to calculate the marginal contribution between agent and mutually agreeable division of cost for IDS deployment. IDS agents are grouped into coalition groups by the various threat levels so as to provide fair and optimal IDS deployment.

A numerical example of the proposed framework is used to investigate the SRV agent and to verify that the Shapley value does indeed obtain the quantitative value for decision and analysis of the IDS. With four threat levels, the different agents' SRVs interact with each other and form coalitions for investigating the most significant agents in various situations. Since the different threat levels indicate the various strengths of attack, a fair and equitable Shapley value describes a stable strategy; the proposed framework can then be used to manipulate the scale to prioritize economical IDS agents. Given limitations due to cost, the administrator can easily deploy the optimal IDS agents for detecting attacks.

The existing IDS overlay network is described in the next section. In Section 3, we provide a framework which consists of a security risk game and a deployment agent game for distributed IDS overlay network. Section 4 presents numerical examples to evaluate the proposed framework and briefly discusses the observed experimental results. In Section 5, we survey related work using the game theoretic approach for network security. In Section 6, some conclusions and directions for future research are given.

2 IDS Overlay Network

The deployment of the critical IDS agent is supported by the information security operations center (SOC) which provides the central management and monitoring of a large-scale network. For example, in Fig. 1, assume that the intruder would choose a path from the network to deliver a malicious packet from Host a (attacker)

Fig. 1 Deployment of the
minimum set of agents in the
IDS overlay network

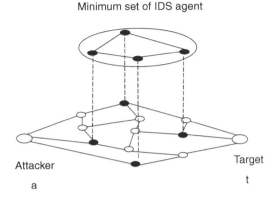

Minimum set of IDS agent

Attacker

a

Target

t

to Host t (target), and a distributed IDS provides specific nodes, $N = \{n_1, n_2, \ldots, n_N\}$, where a node is defined as an autonomous software (agent) which resides at a host (e.g., router, switches, firewall, server, or station), and a host can hold only one IDS agent. The IDS agent utilizes specific techniques, such as signature comparison, pattern detection, and statistical analysis so as to report and prevent possible intrusions or anomalies occurring in a secure network. The IDS administrator arranges for certain detecting agents (nodes) to monitor and control the security of the entire common enterprise network. These detecting agents are mutually influenced by different security measures, such as the intrusion detection rate, false alarm rate, attacker bandwidth consumption, and so on.

The proposed game theoretic framework provides an optimal set of IDS agents to be deployed. The agent's SRV is utilized to create the security overlay network. The SRV of the agent is degree of the attack that is derived from previously defined security measures. In this study we deploy a mini-set of IDS detection agents in the IDS overlay network, which maximize the prevention of network attacks and minimize the number of current resource nodes; see Fig. 1. We also hypothesize each IDS agent relative to the others for specific warning situations in the secure overlay network.

3 DAMAG

In this study two games are constructed, which represent the two stages needed for economical deployment of IDS. In the first step, we model interactions between network attackers and IDS agent behavior within a two-person, nonzero-sum, non-cooperative game. Moreover, we introduce security measures (i.e., intrusion detection rate, agent migration rate, false alarm rate, and attacker bandwidth consumption). Then the proposed security risk game generates NE as the unique SRV of an agent. In the first step the security risk game is applied and in the second step the deployment agent game is applied. In the first step, the average threshold of

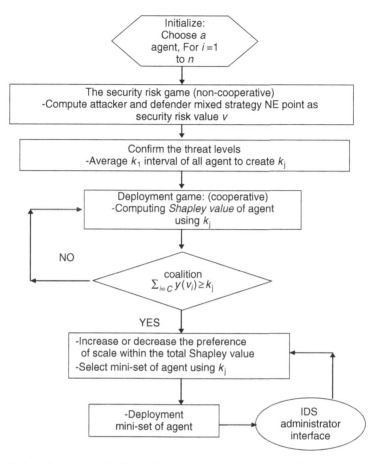

Fig. 2 Optimal deployment algorithm using the two-game theoretic models

all agents is computed in order to confirm the threat level. In the second step, the SRV of each IDS agent obtained from the first step is utilized. In order to establish fair deployment, we construct an IDS agent for a cooperative game theoretic model. A simplified flow chart and algorithm describing the principles of optimal IDS deployment is given in Fig. 2. The foremost goal of the second step is simplicity and scalability. Using the Shapley value as the power index we can create fair and efficient IDS deployment for a secure overlay network which consistently produces an optimal allocation of equitable detection node security values so as to help the system administrator make decisions. Finally, the last procedure increases or decreases the given scale for deploying the appropriate IDS agent in the secure overlay network within the various threat levels (i.e., thresholds).

3.1 The Security Risk Game

In general, the security risk may be defined in terms of the frequency with which breaches are expected to occur (or the security breach rate). Security breaches in an information system result when an attacker succeeds in violating a desired security property. It is measured as the expected cost of loss due to a potential security risk event [19, 20] and multiple security breaches.

$$\text{Security risk} = (\text{security breach rate}) \times (\text{average cost per breach})$$

For any given threat, the IDS is only as strong as it is difficult for the attacker to succeed in the attack. Security strength metrics exist to quantify the detection rate, effort, and other resources required to bypass a system's safeguards against attack. In order to measure the strength of IDS security, we estimate the security breach rate (i.e., attacker's and agent's parameters) and the cost per breach. Thus, the aim of the first proposed model is to obtain the SRV of IDS agent that captures an attacker and the IDS agent behaviors within a two-person game theoretic model (as security strength metrics). Since security strength metrics survey attack threats from the perspective of the adversary, they complement security strength metrics which measure security from the perspective of those the IDS agents are intended to defend. In this chapter, we assume that the attacker and the IDS agent are rational players. Their security measures are defined as following:

Attacker. The population of users in the proposed framework of the network is monitored by the IDS. The user may become an attacker because of curiosity, to achieve self-respect or peer approval, the desire to intentionally destroy, etc. [21]. Thus, we present associated attacked measures. The attacker pays the cost, b_2, which means he gains a profit if intrusion is undetected. However, if the intrusion is detected, the intruder (attacker) pays a punitive cost, b_1. We also assume that $(b_1 - b_2) \leq 0$; that is, an attacker who is detected gains a positive utility, while if $(b_1 - b_2) \geq 0$ he pays a negative utility. We define scattered attacks as penetration past the IDS agent; they can propagate malicious codes rapidly and widely, such as was the case with DDoS and the code red worm [24]. In contrast, the pure attack is a single event, which does not propagate viruses or malicious codes. When the attacker starts the propagation attack on the ith agent and the sources of the DDoS or malicious code from the agent's host, its propagation rate is l_i. Moreover, the ease of user access to the network tends to attract attacks. f_i denotes the network bandwidth of access rate for the ith agent residing at the network host (to demonstrate the influence between attacker and agent). Alpcan and Basar [1] pointed out a based trade-off between security risk and ease of access network. The less a network host is protected by agent's detection mechanisms, the less difficult it becomes to access it, hence more convenient, and vice versa. So, if f_i increases, the probability of attack increases. On the other hand, if f_i decreases, the probability of attack decreases. The propagation attack presents λf_i as multiple of pure attack f_i.

IDS agent. The set of IDS agents is denoted as $N = \{n_1, n_2, \ldots, n_N\}$. An agent resides at the IDS agent's host which refers to the network entity (e.g., router, switches, firewall, server, or station). The goal of the IDS agent, after investigation of the user's transaction, is to classify the user as a normal user or an attacker. Keromytis [8] proposed using an SOS architecture that requires the nodes to frequently **randomize** its location **into the forwarding architecture**, so as to prevent the attack. Thus m_i denotes the migration rate (mobility) of agent n_i at the host where the IDS agent moves to another host. The agent detection scheme commonly shows two types of errors: the classification of an attacker as a normal user (false negative or missing attack) and the classification of a normal user as an attacker (false positive or false alarm). A curve of receiver operating characteristics (ROC) is used for characterization. IDS configures to operate at a higher detection rate will result in a higher false alarm rate. The ROC curve of IDS also presents the trade-off between false alarm rate and detection rate [5]. Thus, for each agent's n_i protective measures parameters there is an intrusion detection rate p_d, false-negative rate ($1-p_d$) and false alarm rate p_f. The cost value $-c_1$ is the gain of the IDS agent for the detection of an attack while c_2 and c_3 are the costs paid by the IDS agent for false alarms and false-negative rate, respectively.

A normal (or strategic) form represents the attacker and the IDS agent's game, which includes the set of players: {Attacker, IDS agent}, and the sets of strategies: $S_{attacker} = \{u_1, u_2, u_3\}$; $S_{agent} = \{d_1, d_2\}$. The u_1 denotes the strategy of pure attack by the attacker with a probability of r_1, the u_2 denotes the strategy of propagation attack with a probability of r_2, and the u_3 denotes that the attacker does nothing and the probability is $1-r_1-r_2$. The d_1 denotes the strategy of the IDS agent which consists of detecting the attack and the alarm response. The d_2 denotes the strategy of the IDS agent, to do nothing following the attack. In addition, let q and $1-q$ be the probability of detection d_1 and no detection d_2, by the agent, respectively. Therefore, the security risk game is shown in Table 1. The payoff of security risk game is expanded by adding a row and a column that represents a mixed strategy. The mixed strategy of the agent is called q-mix and that of the attacker is called r-mix [6].

Table 1 The security risk game

| Attack | IDS Agent | | |
	d_1	d_2	q-mix
u_1	$b_1 f_i, -c_1(1+p_d+m_i)$	$-b_2 f_i, c_3(1+(1-p_d))$	$b_1 f_i q - b_2 f_i (1-q)$
u_2	$b_1(1+\lambda f_i+l_i),$ $-c_1(1+p_d+l_i+m_i)$	$-b_2(1+\lambda f_i+l_i),$ $c_3(1+l_i+(1-p_d))$	$b_1(1+\lambda f_i+l_i)q$ $-b_2(1+\lambda f_i+l_i)(1-q)$
u_3	$0, c_2(1+p_f+m_i)$	$0, 0$	0
r-mix	$-c_1(1+p_d+m_i)r_1$ $-c_1(1+p_d+l_i+m_i)r_2$ $+c_2(1+p_f+m_i)(1-r_1-r_2)$	$c_3(1+(1-p_d))r_1$ $+c_3(1+l_i+(1-p_d))r_2$	

Similar to the model of Alpcan and Basar [1], our game has no pure strategy equilibrium, so the analysis is extended by considering mixed strategies with the player probability distribution in the space of their pure strategies [2]. The Nash equilibrium presents a steady state in a game in which the players' choices are not deterministic but are regulated by probabilistic rules [13]. The player has a mixed strategy NE; each player only considers the average payoff function of his coplayer, rather than optimizing his own average cost function [6].

Every normal-form game has at least one Nash equilibrium in mixed strategies. In the next step the mixed strategy Nash equilibrium, given by 3×2 bi-matrix, is computed. The strategy pair (r^*, q^*) is said to constitute a non-cooperative mixed NE to the security risk game if the following equivalent inequalities are satisfied [2]:

$$
\begin{aligned}
& q^* \left[(-c_1 (1 + p_d + m_i) r^*_1 - c_1 (1 + p_d + l_i + m_i) r^*_2 \right. \\
& \quad + c_2 (1 + p_f + m_i) (1 - r^*_1 - r^*_2))] \\
& \quad + (1 - q^*) \left[(c_3 (1 + (1 - p_d)) r^*_1 + c_3 (1 + l_i + (1 - p_d)) r^*_2) \right] \\
& \leq \cdot q \left[(-c_1 (1 + p_d + m_i) r^*_1 - c_1 (1 + p_d + l_i + m_i) r^*_2 \right. \\
& \quad + c_2 (1 + p_f + m_i) (1 - r^*_1 - r^*_2))] \\
& \quad + (1 - q^*) \left[- (c_3 (1 + (1 - p_d)) r^*_1 + c_3 (1 + l_i + (1 - p_d)) r^*_2) \right], \\
& r^* \left[b_1 f_i q^* - b_2 f_i (1 - q^*) \right] + (1 - r^*) \left[b_1 (1 + \lambda f_i + l_i) q^* \right. \\
& \quad - b_2 (1 + \lambda f_i + l_i) (1 - q^*)] \\
& \leq \cdot r [b_1 f_i q^* - b_2 f_i (1 - q^*) + (1 - r^*) \left[b_1 (1 + \lambda f_i + l_i) q^* \right. \\
& \quad - b_2 (1 + \lambda f_i + l_i) (1 - q^*)], \\
& r^* \left[b_1 f_i q^* - b_2 f_i (1 - q^*) \right] \leq r \left[b_1 f_i q^* - b_2 f_i (1 - q^*) \right], \\
& r^* \left[b_1 (1 + \lambda f_i + l_i) q^* - b_2 (1 + \lambda f_i + l_i) (1 - q^*) \right] \\
& \leq r \left[b_1 (1 + \lambda f_i + l_i) q^* - b_2 (1 + \lambda f_i + l_i)(1 - q^*) \right]
\end{aligned} \tag{1}
$$

where $0 \leq r \, q \leq 1$. The Eq. 1 solves the mixed strategy Nash equilibrium pair (r^*, q^*) which is the optimal for attacking and detection.

The attacker has three strategies and must consider the three different actions in her mix. IDS agent also has two strategies. So, they consist of the three situations. We can use the prevent-exploitation method [6] that finds r^* and q^* from these strategies' intersections in the following three NE pairs.

1. If $-c_1(1+p_d+m_i)r-c_1(1+p_d+l_i+m_i)(1-r) = c_3(1+(1-p_d))r+c_3(1+l_i+(1-p_d))(1-r)$
    and $b_1 f_i q - b_2 f_i(1-q) = b_1(1+\lambda f_i+l_i)q - b_2(1+\lambda f_i+l_i)(1-q)$ then

$$
\begin{aligned}
r^* &= \frac{c_1 (1 + p_d + l_i + m_i) + c_3(1 + l_i + (1 - p_d))}{-c_1(1 + p_d + m_i) + c_1(1 + p_d + l_i + m_i) - c_3(1 + l_i + (1 - p_d)) + c_3(1 + (1 - p_d))} \\
q^* &= \frac{-b_2 (1 + \lambda f_i + l_i) + b_2 f_i}{b_1 (1 + \lambda f_i + l_i) + b_2 (1 + \lambda f_i + l_i) - (b_1 + b_2) f_i}
\end{aligned} \tag{2}
$$

2. If $-c_1(1+p_d+m_i)r+c_2(1+p_f+m_i)(1-r) = c_3(1+l_i+(1-p_d))r$ and $b_1 f_i q - b_2 f_i(1-q^*)=0$
    then

$$r^* = \frac{c_2 \left(1 + p_f + m_i\right)}{c_1 \left(1 + p_d + m_i\right) + c_2 \left(1 + p_f + m_i\right) + c_3 \left(1 + l_i + (1 - p_d)\right)} \tag{3}$$

$$q^* = \frac{b_2 f_i}{(b_1 + b_2) f_i}$$

3. If $-c_1(1+p_d+l_i+m_i)r+c_2(1+p_f+m_i)(1-r)=c_3(1+l_i+(1-p_d))r$ and $b_1(1+\lambda f_i+l_i)$ $q-b_2(1+\lambda f_i+l_i)(1-q) = 0$ then

$$r^* = \frac{c_2 \left(1 + p_f + m_i\right)}{c_1 \left(1 + p_d + l_i + m_i\right) + c_2 \left(1 + p_f + m_i\right) + c_3 \left(1 + l_i + (1 - p_d)\right)} \tag{4}$$

$$q^* = \frac{b_2 \left(1 + \lambda f_i + l_i\right)}{b_1 \left(1 + \lambda f_i + l_i\right) + b_2 \left(1 + \lambda f_i + l_i\right)}$$

A mixed NE pair (r^*, q^*) is an optimal strategy which presents the probability vector $r^*_i = \{r^*(u_1), r^*(u_2), r^*(u_3)\}$ with actions $\{u_1, u_2, u_3\}$ of the attacker and the vector $q^* = \{q^*(d_1), q^*(d_2)\}$ with actions $\{d_1, d_2\}$ of the IDS agent. Let v_i be the ith agent's SRV, which considers detection and attack probabilities. It can be computed by Equation (5).

$$v_i = \frac{r^*_i(u_1) + r^*_i(u_2)}{r^*_i(u_3)} + \frac{q^*_i(d_1)}{q^*_i(d_2)} \quad i \in N \tag{5}$$

where $r^*(u_1)$ and $r^*(u_2)$ indicate the attacker in optimal mixed strategy probabilities for pure and proliferated attacks, respectively.

The $q^*(d_1)$ and $q^*(d_2)$ denote the optimal mixed strategy for the ith agent detection and no detection, respectively. A mixed strategy equilibrium predicts that the outcome of a game is stochastic, and it can capture stochastic regularity [13]. A mixed strategy Nash equilibrium is similar to a stochastic state. The attacker and agent have information about their payoffs with which actions were taken in the past; each player applies these payoffs to form his belief about the future behavior of the other players, and hence formulate his optimal mixed strategy. Therefore, v_i derived from two (the attacker and agent) optimal strategy probabilities for the ith agent presents the security risk value. The next proposed model uses SRV to compute the Shapley value of each agent within the cooperative game.

3.2 The Agent Deployment Game

In this section, we liken the interaction of IDS agent to the playing of a cooperative game in the IDS overlay network. A fair and efficient method is needed for deciding the number of and prioritizing the agents to be deployed for detection. The Shapley value is a power index for cost allocation [16]. The cooperative game provides a

suitable model for the design and analysis of detection agent deployment, and it is
shown that the famous Shapley value rule satisfies many nice fairness properties
[23]. Thus, in the proposed model the Shapley value is applied to create optimal
cost allocation for IDS deployment.

We define $y: V \rightarrow R^+$ as a one-to-one function by assigning a positive real
number to each element of v and $y(0) = 0$, $V = \{v_1, v_2,\ldots,v_j\}$, $j\in n$. The security of
IDS deployment is based on the concept of the threat level, h. The security threat
levels, $H = \{h_1,\ldots, h_H\}$, where $0 < k_1 < k_2 < \ldots < k_H$ are the corresponding threshold
values. In Equation (6), given the agent's output vector v, the security level, L, of
the IDS is equal to h_j, if the sum of the SRVs of the agents is greater than or equal
to k_j:

$$
L = \begin{cases}
h_1 & \text{if } \sum_{i=1}^{N} y(v_i) \geq k_1 \\
h_j & \text{if } \sum_{i=1}^{N} y(v_i) \geq k_j \\
h_{j+1} & \text{if } \sum_{i=1}^{N} y(v_i) \geq k_{j+1} \\
h_H & \text{if } \sum_{i=1}^{N} y(v_i) \geq k_H
\end{cases}
\tag{6}
$$

where $k_1 = v_{\text{Mini}} + k_{\text{in}}$, $k_j = v_{\text{Mini}} + jk_{\text{in}}$, $k_{j+1} = v_{\text{Mini}} + (j+1)k_{\text{in}},\ldots, k_H = v_{\text{Mini}} + Hk_{\text{in}}$

$$
k_{\text{in}} = \left(\frac{v_{\text{Max}} - v_{\text{Mini}}}{H + 1} \right)
\tag{7}
$$

The agents can be grouped into different security levels subject to the value of
k_j of the threshold. It is divided by $H+1$ threat levels from maximum SRV v_{Max} to
minimum v_{Mini}. So this method divided the security levels averagely.

The SRV of the agent can be modeled as an N-person game with $X = \{1, 2,\ldots,$
$N\}$, which includes the set of players (i.e., agent) and each subset $V \subset N$, and
where $v_j \neq 0 \ \forall_j \in V$ is called a coalition [6, 10]. The coalition of X agent groups
in the kth threshold of the threat level, and each subset of X (coalition), represents
the observed threat pattern for different threat levels H. The aggregate value of the
coalition is defined as the sum of the SRVs of the agent, $y(C) = \sum_{i\in C} y(v_i)$, and
is called a coalition function. Each agent coincides with the given one or another of
the k thresholds of the threat level. The different priorities for agent deployment can
be derived from the various thresholds. According to intrusion threat for each IDS
host with respect to others and the effect of the threshold values on various threat
levels, Shapley value represents the relative importance of each agent. Let $y(C) =
\sum_{i\in C} y(v_i)$, $v_i \in V$, $C \subset X$ be the value of the coalition C with cardinality c. Then, the
Shapley value of the ith element of the agent vector is defined by

$$
S(i): = \sum_{\substack{C \subset X \\ i \in C}} \frac{(c-1)!(n-c)!}{n!} \left[y(C) - y(C - \{i\}) \right]
\tag{8}
$$

In determining whether the SRV of the ith agent is greater than or equal to the value of the threshold of the hth threat level (i.e., coalition), the formula can be simplified to

$$S(i): = \sum_{C' \subset X} \frac{(c-1)!(n-c)!}{n!} \qquad (9)$$

Equation (8) can be simplified to Equation (9) because the term $y(C)–y(C–\{i\})$ will always have a value of 0 or 1, taking the value 1 whenever C' is a winning coalition but not $C^-\{i\}$ [17]. Hence, Shapley value is $S(i)$ where C' denotes the winning coalitions with $\sum_{i \in C'} y(v_i) \ge k_j$. The Shapley value of the ith agent output indicates the relative SRV for the various thresholds (i.e., threat levels) so as to provide a computation for choosing a reasonable agent for IDS deployment.

In Fig. 3, we input the set of agent SRVs to compute the vector of the Shapley value when considering the various thresholds k_j. We assume the threat level to the IDS be a kind of security situation in which the network faces attacks. DAMAG computes the aggregation of the agent's SRV and presents the various agent coalitions in the overlay network according to the various threat levels. Let $P^h = \{n_1^h, n_2^h, n_3^h, \ldots, n_i^h\}$, $\forall_i \in N$, be the priority of the agent set, which is subject to the threat level h (e.g., green, yellow, orange, or red levels). Therefore, administrator decides which IDS agents are to resist the attacks based on the different security situations. This rating system is based entirely on the estimation of the utility's desirability for deployment of the IDS agent. The administrator can also utilize an adjustable scale to make decisions, which would select an optimal set of detection agents for resistance of attacks.

Fig. 3 Deployment priority of the IDS agent set obtained by using the *Shapley value*

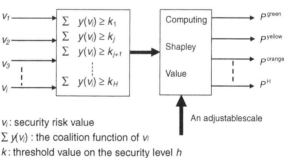

v_i: security risk value
$\sum y(v_i)$: the coalition function of v_i
k: threshold value on the security level h
P: the priority of agent set

4 Simulation Experiments

Numerical examples to verify the application of the proposed framework are presented in this section. First, 20 hypothetical number of parameters is given to model the bi-matrix, and randomly generate simulation sets of security measures for agents and attackers in Table 2. Then, we apply the GAMBIT [14] to calculate 20 agents'

Table 2. Numerical examples for the generation of SRVs

Agent	Attacker and agent's parameters of moves										NE
	b_1	$-b_2$	$f_i\%$	$-c_1$	c_2	c_3	$l_i\%$	$m_i\%$	$p_d\%$	$p_f\%$	v_i
n_1	10	−100	0.49	−30	80	60	0.8	0.3	0.5	0.5	10.913
n_2	20	−90	0.48	−90	20	10	0.2	0.9	0.9	0.2	4.66
n_3	30	−80	0.4	−80	30	30	0.3	0.8	0.8	0.3	2.925
n_4	40	−70	0.35	−70	40	50	0.4	0.7	0.7	0.4	2.11
n_5	50	−60	0.3	−60	50	60	0.5	0.6	0.6	0.5	1.685
n_6	60	−50	0.25	−50	60	40	0.6	0.5	0.8	0.6	1.589
n_7	70	−40	0.2	−40	70	80	0.7	0.4	0.9	0.4	1.264
n_8	80	−30	0.15	−30	80	60	0.8	0.3	0.8	0.3	1.283
n_9	90	−20	0.1	−20	90	80	0.9	0.2	0.6	0.1	1.009
n_{10}	100	−10	0.05	−10	100	90	1	0.1	0.5	0.2	0.96
n_{11}	5	−95	0.48	−1	99	98	0.05	0.99	0.5	0.1	20.367
n_{12}	15	−85	0.44	−85	15	14	0.15	0.85	0.5	0.1	5.799
n_{13}	2	−99	0.38	−2	98	100	0.99	0.99	0.5	0.1	50.638
n_{14}	25	−55	0.32	−65	35	30	0.35	0.65	0.7	0.2	2.538
n_{15}	100	−10	0.05	−1	100	90	1	1	0.5	0.3	1.497
n_{16}	55	−45	0.22	−45	55	50	0.55	0.45	0.8	0.45	1.461
n_{17}	10	−100	0.49	−30	80	85	0.8	0.3	0.5	0.3	10.675
n_{18}	75	−25	0.12	−25	75	80	0.75	0.25	0.8	0.25	1.093
n_{19}	90	−20	0.1	−85	15	16	0.15	0.85	0.5	0.5	0.38
n_{20}	100	−10	0.05	−90	20	22	0.2	0.9	0.9	0.2	0.252

$\lambda = 2$

SRVs from the numerical examples and create one Nash equilibrium by Logit tracing in extensive game. Second, the SRV of each IDS agent is utilized for computing each agent's Shapley value. In this chapter, four threat levels are designed for the distributed IDS (i.e., green, yellow, orange, and red). We adopt the Matlab to find the Shapley value of agents for the four threat levels. Then simulations create four agent deployment sets for a secure IDS network.

We demonstrate 20 experimental agents in a secure IDS overlay network. The 20 SRVs are calculated exactly using Equations (2)–(4). Figure 4 (a) shows the sequence of 20 agent SRV vector outputs, $V = [v_{20}, v_{19}, v_{10}, v_9, v_{18}, v_7, v_8, v_{16}, v_{15}, v_6, v_5, v_4, v_{14}, v_3, v_2, v_{12}, v_{17}, v_1, v_{11}, v_{13}] = [0.252, 0.38, 0.96, 1.009, 1.093, 1.264, 1.283, 1.461, 1.497, 1.589, 1.685, 2.11, 2.538, 2.925, 4.66, 5.799, 10.675, 10.913, 20.367, 50.638]$. We use Equation (7) to obtain four thresholds $k_{\text{green}} = 10.08$, $k_{\text{yellow}} = 20.15$, $k_{\text{orange}} = 30.23$, $k_{\text{red}} = 40.31$ according to the SRV vector output. Using these threshold values we apply Equation (9) to calculate the exact Shapley value of IDS detection agents as shown in Fig. 4(b). The threat level approach proposed (shown in Fig. 4(b)) is a more objective and efficient deployment than pure SRV (shown in Fig. 4(a)). It is interesting to note that the Shapley value of IDS agents with a low SRV is smaller for the red and orange threat levels than the ones for the yellow and green threat levels. This indicates that agents with low SRVs play a less significant role than others for switching to the red threat advisory.

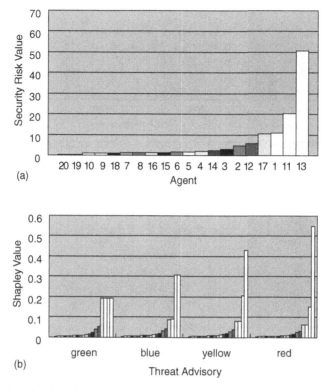

(a)

(b)

Fig. 4 Security value risk of the 20 IDS agents (**a**) and considering the four threat levels (**b**)

Figure 4(b) also illustrates how the high SRV agents play a more significant role than do the lower ones in dangerous situations (e.g., red level) considering four threat levels. On the other hand, there is no particular difference between high and low SRV agents in gentle dangerous situations (e.g., green level). So, the proposed framework can also provide more quantitative values for various situations than in human decision-making, as is illustrated from a comparison of Fig. 4(a) with (b).

We also apply an adjustable preference scale so as to build a rating approach for the IDS deployment which provides a fair and equitable cost allocation. Administrator can adjust this preference scale to agent deployment of IDS within the four threat levels because of resource limitation. The fairness of the Shapley value for three experimental cases is examined by the proposed preference scale. Moreover, we respectively compare each agent's Shapley value (bigger than 0.1, 0.05, and 0.01) within the four minimum sets, indicated in Fig. 5 (a–c).

First, the preference scale of Shapley value is bigger than 0.01. Figure 5(a) indicates that 12 agents $P^{green} = \{n_{15}, n_6, n_5, n_4, n_{14}, n_3, n_2, n_{12}, n_{17}, n_1, n_{11}, n_{13}\}$ are deployed in the IDS overlay network when the threat level is green. Eleven agents $P^{yellow} = P^{orange} = \{n_6, n_5, n_4, n_{14}, n_3, n_2, n_{12}, n_{17}, n_1, n_{11}, n_{13}\}$ are deployed in

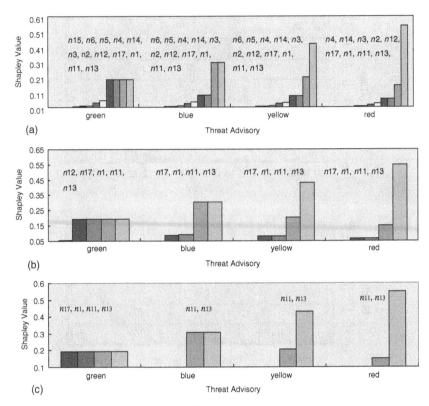

Fig. 5 Three experimental deployments of IDS for four threat advisory levels. (**a**) Shapley value $S(i) > 0.7$. (**b**) Shapley value $S(i) > 0.8$. (**c**) Shapley value $S(i) > 0.9$

the IDS overlay network when the threat levels are yellow and orange. Nine agents $P^{\text{red}} = \{n_4, n_{14}, n_3, n_2, n_{12}, n_{17}, n_1, n_{11}, n_{13}\}$ are deployed for the red threat level.

Second, the preference scale of Shapley value is bigger than 0.05. Figure 5(b) indicates that five agents $P^{\text{green}} = \{n_{12}, n_{17}, n_1, n_{11}, n_{13}\}$ are deployed in the IDS overlay network when the threat level is green. Four agents $P^{\text{yellow}} = P^{\text{orange}} = P^{\text{red}} = \{n_{17}, n_1, n_{11}, n_{13}\}$ are deployed for the yellow, orange, and red threat levels.

Third, the preference scale of Shapley value is bigger than 0.1. Figure 5(c) indicates that four agents $P^{\text{green}} = \{n_{17}, n_1, n_{11}, n_{13}\}$ are deployed in the IDS overlay network when the threat level is green, and two agents $P^{\text{yellow}} = P^{\text{orange}} = P^{\text{red}} = \{n_{11}, n_{13}\}$ are deployed for the yellow, orange, and red threat levels.

Figure 5(a–c) shows that different threat levels (advisories) efficiently provide specific selection of IDS agent to the administrator when the preference scale is adjusted. Figure 5(c) shows that small numbers of higher SRV value agents are paid by specific resources so as to control few serious attacks suddenly and neglect to others if administrator changes the preference scale to serious situations (e.g., red level). In contrast, Fig. 5(a) shows that most of agents are averagely paid by specific resources so as to find out attacks gradually and prevent them from propagation if administrator changes the preference scale to gentle dangerous situations (e.g., green level). Figure 5(a–c) shows that it provides the administrator with a way to prioritize IDS agents when detecting attacks for the different threat levels. Therefore, the proposed framework deploys the fair agent of IDS for threat advisory.

5 Related Works

Various game theoretic approaches have been applied to model the interactions between attackers and defenders for computer network intrusion detection. Burke [3] proposed a two-player game framework to model information warfare between two players, an attacker and a system administrator. In their model the mixed strategy equilibrium, which constructs a player's optimal solution, was used. The strategies and scenarios of the model are simple extensions with a few simulations to analyze how it would behave in real situations.

Lye and Wing presented a zero-sum, stochastic game model which included two players' moves and changes of state process [12]. Changes of state between attacker and protection are analyzed using Markov's decision process in order to compute the optimal solution for enhancing the security network and tracing the attacker. They are concerned that the more moves the two players make, the greater the lack of scalability and extension with the repeated game.

Liu et al. proposed a multi-stage game that could automatically infer the attacker and the system defense strategy [10]. They developed formalization of inference for the evaluation and analysis of the attacker's intention, objectives, and strategies, and compute the NE of the game to infer what would be optimal moves for the security system to counter the attackers. They also presented a multi-stage, signaling game for detecting attacks behaviors and for predicting the actions of attackers [11]. A credit card fraud scheme was modeled to show the eventual real effects of attacker actions.

Kodialam and Lakshman proposed using a zero-sum model to develop a network packet sampling strategy yielding the NE for improving the probability of detecting attacks [9]. While their model utilized more in-depth mathematical concepts than did the previous model, they still suffered from scalability and extension of the computed equilibrium solution.

Cavusoglu addresses the moves of the hacker and the system as an inspection game similar to arms control and disarmament or environmental control games [5]. They analyze the value of the IDSs using backward induction from the components of the model (i.e., user, firm, and technology). Their model has the drawback that the user has only two moves (i.e., to attack or not to attack) which does not meet the real situation of IDS in detection attacks. We study attacks that can propagate malicious codes rapidly and widely (e.g., DDoS, code red worm).

Alpcan and Basar proposed the utilization of cooperative and non-cooperative game theory concepts to address some of the basic network security trade-offs [1]. They constructed a sensor warning system and a security game for various decisions, analysis, and IDS control schemes. The sensor warning system generates a security risk value for the agent. This value could have various levels but the calculation is based on simple detection output. However, the strategy between the IDS and the attacker was incomplete so could not satisfy the decision-making requirements for IDS control. They proposed two independent schemes.

In all the above research, it is proposed that a game theoretic model be utilized for capturing the interaction between the attacker and the defender. Obviously, they meet common problems, related to extension and scalability. We propose a framework for connecting the security risk model and IDS agent deployment models. Since an attacker has three moves and the IDS agent has two moves, these problems do not exist in our framework.

As far as we know, we have not yet found a framework for the fair allocation of limited cost for the deployment of the IDS agent. This is known as a cost allocation problem (CAP), and to decide whether a given allocation is fair is generally an NP-complete problem [7]. Recently, Shapley value division has been used in transmission system and wireless mobile ad hoc networks to allocate system costs [4, 15]. Comparisons of cooperative game solutions with traditional allocation systems verify they are better in economic and physical terms. The proposed Shapley value formula in this chapter is based on Owen's method [17], which is described in detail in Section 3.2. We simplify the Shapley value formula to calculate the accurate value for large numbers according to the formation of a majority coalition of the agent in accordance with the threat level. The various threat levels are derived from the security risk value of each agent; the DAMAG framework also provides recommended protective measures for the security advisory system.

6 Conclusions and Future Work

We investigate existing schemes for resisting DDoS, including reactive prevention and proactive mitigation schemes. Recently, IDS administrators have been faced with a growing choice of security risks requiring a trade-off between ease of network use (ease of access) and the capability of system detection (detection overhead), in their discovery of IDS deployment problems. In this research, we apply game theoretic concepts to address this trade-off between security risk and ease of access for decision-making deployment of IDS overlay networks. The proposed framework is developed for IDS decision and analysis processing. It consists of two stages. The first game theoretic model is used to model the mixed strategy NE of the game to find the security risk value (SRV) of each agent, and the second model utilizes the SRV to compute the Shapley value for each IDS agent given the various threat levels. In order to build up a rating approach for deployment of IDS, we apply an adjustable Shapley value scale, which provides fair and equitable cost allocation. Additionally, we provide numerical examples demonstrating the suitability of game theoretic models to the development of IDS agents. The experiments actually identify and provide encouragement for the use of the framework to connect the Nash equilibrium and Shapley value concepts, enabling the IDS to prioritize deployment for four threat levels or situations. Future work will extend our proposed model and present a simulation using real data obtained from an organization's security operational center.

References

1. Alpcan, T., Basar, T. (2003). A game theoretic approach to decision and analysis in network intrusion detection, IEEE Conference on Decision and Control, 2595–2600.
2. Basar, T., Olsder, G. J. (1999). Dynamic Noncooperative Game Theory, Academic Press, Philadelphia, 87.
3. Burke, D. A. (1999). Towards a game theoretic model of information warfare. Masters Thesis, Air Force Institute of Technology, Air University.
4. Cai, J., Pooch, U. (2004). Allocate fair payoff for cooperation in wireless ad hoc networks using Shapley value, Proceedings of the 18th International Parallel and Distributed Processing Symposium, 219–227.
5. Cavusoglu, H., Mishra, B., Raghunathan, S. (2005). The value of intrusion detection systems in information technology security architecture, Information Systems Research, 16, 28–46.
6. Dixit, A. Skeath, S. (2001). Games of Strategy, W. W. Norton & Company.
7. Goemans, M. X., Skutella, M. (2004). Cooperative facility location games, Journal of Algorithms, 50, 194–214.
8. Keromytis, A. D., Vishal, M., Rubenstein, D. (2004). SOS: An architecture for mitigating DDoS attacks, IEEE Communications, 22, 176–188.
9. Kodialam, M. T., Lakshman, V. (2003). Detecting network intrusions via sampling: a game theoretic approach, INFOCOM, 2003, 1880–1889.
10. Liu, P., Zang, W. (2005). Incentive-based modeling and inference of attacker intent, objectives, and strategies, ACM Transactions on Information and System Security, 8, 78–118.
11. Liu, P., Li, L. (2002). A Game Theoretic Approach to Attack Prediction, Technical Report, Penn State University.
12. Lye, K. W., Wing, J. (2005). Game strategies in network security, International Journal of Information Security, 4, 71–86.
13. Martin, O. J., Ariel, R. (1994). A Course in Game Theory, MIT Press, Cambridge.
14. McKelvey, R. D., Andrew, M. M., and Turocy, T. L. (2007). Gambit: Software Tools for Game Theory, http://econweb.tamu.edu/gambit .
15. Mel, P., Hu, V., Lippmann, R. J., Zissman, H. M. (2002). An Overview of Issues in Testing Intrusion Detection Systems, NIST, Gaithersburg, MD.
16. Mishra, D., Rangarajan, B. (2005). Cost sharing in a job scheduling problem using the Shapley value, Proceedings of the 6th ACM Conference on Electronic Commerce, 232–239.
17. Owen, G. (2001). Game Theory, 3rd ed. Academic Press, New York.
18. Parsons, S., Wooldridge, M. (2002). Game theory and decision theory in multi-agent systems, Autonomous Agents and Multi-Agent Systems, 5, 243–254.
19. Schechter, S. E. (2004). Computer Security Strength and Risk: A Quantitative Approach, PhD Thesis, Harvard University.
20. Schechter, S. E. (2005). Toward econometric models of the security risk from remote attacks, IEEE Security & Privacy, 3(1), 40–44.
21. Shaw, D. S., Post, J. M., Ruby, K. G. (1999). Inside the minds of the insider, Security Management, 43, 34–44.
22. Symantec Corporation, http://www.symantec.com/index.jsp .
23. Zolezzi, J. M., Rudnick, H. (2002). Transmission cost allocation by cooperative games and coalition formation, IEEE Transactions on Power Systems, 41, 1008–1015.
24. Zou, C. C., Gong, W. Towsley, D. (2002). Code red worm propagation modeling and analysis, In: Proceedings of the 9th ACM Symposium, 138–147.

ETKnet: A Distributed Network System for Sharing Event Data and Multi-faceted Knowledge in a Collaboration Federation

Stanley Y.W. Su, Howard W. Beck, Xuelian Xiao, Jeff DePree, Seema Degwekar, Chen Zhou, Minsoo Lee, Carla Thomas, Andrew Coggeshall, and Richard Bostock

Abstract This chapter presents a distributed event-triggered knowledge network (ETKnet) developed for use by government organizations to share not only data and application operations but also knowledge embedded in organizational and inter-organizational policies, regulations, data, and security constraints as well as collaborative processes and operating procedures. A unified knowledge and process specification language has been developed to formally specify multi-faceted human and organizational knowledge in terms of three types of knowledge rules and rule structures. A user-friendly interface is provided for collaborating organizations to define events of interest as well as automated and manual operations, operation structures, knowledge rules, rule structures, and triggers. Through this interface, these organizations can also perform task management, administrative management, configuration management, and ontology management. Events are published in a global registry for browsing, querying, event subscription, and notification. Rules and rule structures are automatically translated into Web services for

Stanley Y.W. Su, Xuelian Xiao, Jeff DePree, Seema Degwekar
Department of Computer Information Science and Engineering, Database System R&D Center, University of Florida, Gainesville, FL 32611, USA
e-mail: {su,xxiao,jdepree,spd}@cise.ufl.edu

Howard W. Beck, Chen Zhou
Department of Agricultural and Biological Engineering, University of Florida, Gainesville, FL 32611, USA
e-mail: {hwb,czhou}@ufl.edu

Minsoo Lee
Department of Computer Science and Engineering, Ewha Womans University, Seoul 120-750, Korea
e-mail: mlee@ewha.ac.kr

Carla Thomas, Andrew Coggeshall, Richard Bostock
Department of Plant Pathology, University of California, Davis, CA 95616, USA
e-mail: {cthomas, acoggeshall,rmbostock}@ucdavis.edu

C.C. Yang et al. (eds.), *Security Informatics*, Annals of Information Systems 9,
DOI 10.1007/978-1-4419-1325-8_8, © Springer Science+Business Media, LLC 2010

discovery and distributed processing in ETKnet. Event data are dots that can be connected dynamically across organizational boundaries through the interoperation of knowledge rules and processes.

Keywords Knowledge management · Distributed system · Collaboration technology · Event-based system · Agriculture homeland security

1 Introduction

Government organizations worldwide are facing complex problems such as illegal immigration, terrorism, disease outbreaks, and natural disasters. Effective resource sharing, collaboration, and coordination among government organizations are needed to solve these and other complex problems. They need to share not only data and application system functions, but also human and organizational knowledge useful for decision support, problem-solving, and activity coordination. The technology of sharing distributed, heterogeneous data has been extensively studied, but an effective way of sharing knowledge among collaborating organizations is still lacking.

In this work, we are interested in capturing, managing, and applying the multifaceted knowledge embedded in organizational and inter-organizational policies, regulations, constraints, processes, and operating procedures. A common means of representing knowledge is to use knowledge rules [18]. Three types of knowledge rules have been found to be useful in many applications [1, 16]: integrity constraints [22], logic-based derivation rules [23], and action-oriented rules [3, 25]. If they are incorporated into a single knowledge specification language and used together in an interoperable fashion, they can provide a very powerful facility for not only knowledge representation, but also knowledge sharing. Furthermore, based on our work in both e-business and e-government applications [5, 7], we have found that it is necessary to use structures of these different types of rules to model complex collaborative policies or processes. Also, a collaborative process or an operating procedure may need to be enacted when some important event occurs. It will be very useful if knowledge rules and processes/procedures can be specified in a unified specification language so that they can interoperate. By interoperability, we mean that rules of different types and processes/procedures can interface with one another seamlessly. For example, when an event occurs, there is data associated with the event (i.e., event data). A logic-based derivation rule defined by one organization may look at the event data and deduce from it some additional data for use by an action-oriented rule of another organization. This action-oriented rule may activate a structure of operations, which then produces some data that need to be verified by a constraint rule of yet another organization. Or, the above rules and operations can be specified by a single organization in a rule structure triggered by the event.

Sharing useful data is also an important aspect of collaboration. Since each government agency takes the utmost care with the security of its database, opening it up in its entirety for access by other organizations is not an option. Apart from the security reason, it is typically not necessary to allow full access to a database in most practical cases. When organizations collaborate, it is usually to solve specific problems. Thus it is important to devise a framework that allows them to share only those data and knowledge pertaining to those problems. Our approach for achieving data and knowledge sharing is to augment an event subscription and notification system with knowledge sharing and processing capabilities. An event is anything of significance to collaborating organizations (e.g., an arrest, a terrorist incident, the detection of a disease, a special state of a database, a signal from a sensor, etc.) that occurs at a particular point in time. Collaborating organizations would obtain only the data that are pertinent to the occurrence of an event (i.e., event data) and process only those knowledge rules and processes/procedures that are "applicable" to the event data. A rule, process, or procedure is applicable to an event, if the objects and attributes specified in the event data form a superset of the input data needed for processing the rule, process, or procedure. Initially, the event data set contains only the data associated directly with the event occurrence. However, as relevant rules and operations are applied on the event data, new data can be generated and old data can be modified, thus making some other distributed knowledge rules and rule structures applicable. Multiple rounds of event data transmission, rule processing, and data aggregation may take place in order to produce all the data that are pertinent to the event occurrence. Thus, an event-triggered data and knowledge sharing system that facilitates event subscription, event notification, delivery of event data, and processing and interoperation of applicable knowledge rules, rule structures and processes/procedures would be ideal for any collaborative federation. It is therefore the goal of our project to research and develop such a system.

An important issue to be addressed is the interoperability of heterogeneous rules. One possible approach is to choose one rule type as a common format and convert the other rule types into this chosen representation [2]. This approach sounds attractive because it only needs a single rule engine to process all the converted rules. However, since different types of rules have significant semantic disparities, converting a rule from one representation to another may lead to loss of information. Another possible approach is to build wrappers around different types of rule engines [15] and provide a common interface to enable these rule engines to exchange the data generated by rules. In our opinion, this approach is not ideal either because it will result in a very complex system that is difficult to operate and maintain. Another issue is the specification and enactment of collaborative processes and operating procedures. Traditionally, business processes and workflow processes are defined by some definition languages such as BPEL [13] and WPDL [26], and processed by some business process execution engine or workflow management system. Typically, these engines and systems do not interoperate with knowledge management systems.

This chapter reports the main extensions made to our earlier system reported in [6, 7]. We have extended our knowledge specification language and user interface to allow integrated specification of knowledge rules and processes/procedures. Operation structures which model processes or operating procedures can be specified in the action or alternative action part of a condition-action–alternative-action rule. Defined rules (including processes/procedures) and rule structures are then automatically translated into code and wrapped as Web services for their discovery, processing, and interoperation in a Web service infrastructure. This approach avoids the use of multiple rule engines and workflow/process management systems, and the problem of rule-to-rule conversions. Additionally, we have added an ontology management system to the system architecture of ETKnet to achieve ontology-enhanced rule sharing.

2 National Plant Diagnostic Network

The collaborative federation that serves as our application domain is the National Plant Diagnostic Network (NPDN [11]). The U.S. Department of Agriculture (USDA) launched a multi-year national project in May 2002 to build NPDN to link plant diagnostic facilities across the United States. This was done to strengthen the homeland security protection of the nation's agriculture by facilitating quick and accurate detection of disease and pest outbreaks in crops. Such outbreaks can occur as foreign pathogens are introduced into the United States either through accidental importation, by wind currents that traverse continents, or by an intentional act of bioterrorism [24]. NPDN has developed a general standard operating procedure (SOP [21]) to combat a pest-of-concern situation, which details the steps to be taken following detection to obtain a confirmed diagnosis of plant and insect samples and alert relevant people and organizations when such a biosecurity event takes place. At present, NPDN organizations have developed their regional operating procedures using the general SOP as the guide. These procedures are, in most cases, carried out manually. There is a need for their automation. In this work, we have implemented the general SOP in terms of distributed events, rules, rule structures, and operation structures. The system presented in this paper is for deployment in the NPDN environment for plant disease and pest detection and management.

3 Architecture of ETKnet

ETKnet has a peer-to-peer server architecture as shown in Fig. 1. All collaborating organizations have identical subsystems installed at their sites. Each site creates and manages its own events, rules, rule structures, triggers, and operations. Their specifications are registered at the host site of a federation.

A user interface is used at each site to define sharable events, rules, rule structures, triggers, automated application system operations, and manual operations. It is

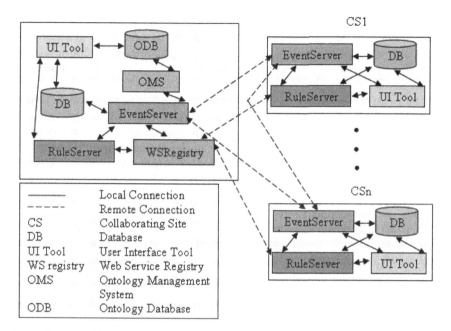

Fig. 1 System architecture

also used for managing tasks assigned to NPDN personnel and the network system. More information about the user interface will be given in Section 5.

ETKnet supports three types of rules: logic-based derivation rule adopted from RuleML [17], constraint rule patterned after [14], and action-oriented rule [25]. Instead of using event–condition–action (ECA) rules, our action-oriented rules have the format: condition-action–alternative-action (CAA). We separate event (E) from CAA rules because they can be defined by different collaborating organizations independently. In both action and alternative action clauses, a single operation or a structure of manual and/or automated operations can be specified. An operation structure can be used to model a complex process or operating procedure. For its specification, we adopt the structural constructs found in BPEL and WPDL, namely sequential, switched, unordered, selective, repeated, And/OR/XOR split, and AND/OR join, which will be explained in Section 5.

For specifying complex organizational policies, constraints and regulations, we use a structure of rules. A rule structure is a directed graph with different types of rules as nodes. These nodes are connected by *link, split, and-join,* and *or-join* constructs. Details about the rule structure will be given in Section 5.

When a shared rule or rule structure is defined, it is automatically translated into program code and wrapped as a Web service. The generated code is stored in a local database and processed by the Rule Server installed at the site where the rule or rule structure is defined. This is for both security and efficiency reasons because the code typically invokes some local application operations. The Web service of the rule or rule structure is registered with the WSRegistry at the host site for discovery and sharing purposes. The integrated specification and processing of knowledge rules and processes/procedures as Web services is significant because it enables their discovery, processing, and interoperation. The algorithm used for translating rules and rules structures into Web services can be found in [7].

The Event Server is responsible for storing information about events defined at that particular site and information about event subscribers. Events defined at all sites are registered with the host site. Users of all organizations can browse or query the registry to subscribe to an event. They can also define triggers to link events to rules and rule structures. In both cases, they are "explicit subscribers." Triggers can be automatically and dynamically generated by the system for those sites that contain applicable rules and rule structures. In that case, the organization that defines the rule or rule structure is an "implicit subscriber" of the event. The subscriber information is automatically downloaded to the Event Server of the site that defines the event for event notification purposes. When an event occurs at a particular site, its Event Server functions as the coordinator of a knowledge sharing session. It carries out event notification by sending event data to both explicit and implicit subscribers to trigger the processing of rules. The Event Server also handles the aggregation of new event data returned from collaborating sites as the result of rule processing. The new version of the event data may trigger another round of rule processing. Multiple rounds of event data transmission and rule processing may take place until no applicable rules exist in the distributed system. At that time, all relevant organizations would have received all the data that are pertinent to the event occurrence.

In a collaborative environment, terms used by different organizations to define events, rules, triggers, and operations can be quite different. It would be beneficial to define a domain-specific ontology to resolve the discrepancies and identify the similarities among specified terms and to facilitate search. Thus, the host site also has an ontology management system including an ontology database. An ontology is a representation of knowledge in a specific domain. It defines a set of concepts within this domain and their relationships. We have defined a domain-specific ontology based on NPDN's standard operating procedure. The ontology database contains the concepts and relationships associated with the event data, terms used in rule and operation specifications, and roles that different people play in NPDN organizations.

The ontology management system we have developed is called Lyra [4]. It consists of a set of tools to create, store, and maintain an ontology. The logic structure of Lyra consists of three layers. The bottom layer is a persistent layer, which stores the ontology as classes, individuals, and properties. The middle layer is an inference layer, which reads the concepts and individuals from the persistent storage, converts them into a normalized form, and performs ontology reasoning which is used for query processing. We enhanced the description logic reasoner, Pellet [20], and used it to perform inference tasks. The top layer is a term layer, which has a structure similar to that of the WordNet [12] lexical database. This layer maps the terms used in users' queries as well as terms used in the specifications of events, rules, and operations to the objects defined in the ontology, so that the discrepancies and similarities between the specified terms can be resolved and identified. We have also developed an ontology editing tool for creating and editing terms and concepts, a rule matcher to facilitate the user's querying for sharable rules, and an application interface for the Rule Server and the Event Server to query for people who play specific roles in NPDN organizations.

4 Distributed Event and Rule Processing

In this section, we use the application in the NPDN domain to explain our distributed event and rule processing technique. We pick a subset of the communication path outlined in the general SOP mentioned in Section 2 and use it as an example.

Figure 2 depicts the following scenario. When a suspect sample is submitted to NPDN Triage Lab, an event, presumptive-positive-sample-observed, is said to have occurred. In the figure, we label this event as E1 for brevity. The occurrence of E1 is denoted as step

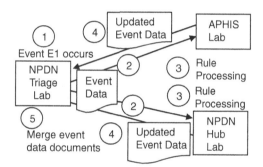

Fig. 2 Distributed event and rule processing

1. E1 at NPDN Triage Lab causes the event data containing the sample information to be sent to an APHIS Confirming Lab and/or NPDN Regional Hub Lab (step 2). This event data is sent to each site as an XML document. These labs have applicable rules that can make use of the event data to provide some more relevant information. APHIS performs confirming diagnosis on the sample (step 3). NPDN Regional Hub Lab informs the appropriate personnel of the sample status (also step 3). These procedures at NPDN Triage Lab and NPDN Regional Hub Lab are modeled using heterogeneous rules and rule structures. Details of these procedures are given later in this section.

The invocation of the applicable rules and rule structures in both the APHIS Lab and NPDN Regional Hub Lab may produce new data or modify the existing data. The new data and updates are then sent back to NPDN Triage Lab as modifications to the original event data document (step 4) and are merged with the original event data (step 5). The new version of event data is sent (not shown in the figure) to APHIS (confirming) Lab and NPDN Regional Hub Lab to begin a second round of rule processing if there are applicable rules. Thus, multiple rounds of event data transmission, rule processing, and data aggregation can take place as event data is dynamically changed by rules. The algorithms for aggregating event data to produce a new version of event data, and for avoiding cyclic processing of distributed rules can be found in [7].

Figure 3 describes the rule structure executed at NPDN Triage Lab upon receiving a sample. This is a structure with both link and split relationships. The first rule, NTLR1, is concerned with acknowledging the receipt of the sample to the sample submitting entity, which can be another diagnostic lab or an independent grower. It asks the lab diagnostician to perform a preliminary diagnosis on the sample. Rule NTLR2 checks if the lab has a Web or distance diagnosis capability. If so, the NPDN Triage Lab is contacted to perform the distance diagnosis. Otherwise, a photograph of the sample is e-mailed. Rule NTLR3 instructs the staff on how to divide and ship the sample to the other labs for a confirming diagnosis. If it is a routine sample, or the Hub Lab has provisional approval to perform the confirming diagnosis, the sample is to be sent to NPDN Hub Lab; otherwise, it is to be sent to a confirming diagnosis designate (CDD, usually an APHIS lab). Rule NTLR4 instructs the staff to contact the campus safety officer to appraise him/her of a presumptive positive sample in the system.

Fig. 3 Rule structures (NPDN Triage Lab, NPDN Hub labs)

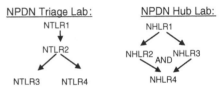

Figure 3 also describes the rule structure executed at NPDN Regional Hub Lab. This lab is responsible for engaging a local expert, if available, to perform the diagnosis and reporting the result back to Triage Lab. The first rule, NHLR1, acknowledges the receipt of the presumptive positive sample sent by NPDN Triage Lab. It also checks to see if proper sample shipping procedures were followed and informs the administrator if otherwise. Rule NHLR2 is concerned with asking a local expert to perform some preliminary diagnosis on the sample. Rule NHLR3, which is processed in parallel with NHLR2, contacts other personnel such as a State Plant Regulatory Official and a State Plant Health Director to inform them of the presumptive positive sample in the system. It is also concerned with determining whether or not Regional Hub Lab needs to send the sample to APHIS Lab for further confirmations. After receiving the diagnosis from the local expert, if APHIS Lab requires Regional Hub Lab to send the sample, rule NHLR4 ensures that the sample is sent. For space reasons, we do not include the rules and rule structures at APHIS Lab.

Some operations in the rules described above are manual operations. They need to be done by a lab staff, and cannot be automated. One example is the process of diagnosing a sample. Our approach to incorporating manual operations is to require all such operations to be defined and registered with the system prior to them being used in a rule. During the registration process, it is necessary to specify either a specific agent who should carry out a manual operation along with his/her contact information, which may include either e-mail or cell phone or both, or a group of agents who play the same role. In the former case, when a rule with a manual operation begins execution, the agent is contacted using either e-mail or text message sent to his/her cell phone, or both, with a message that tells him/her what operation to perform. The instruction also includes how to let ETKnet know when the operation has been completed. In the latter case, all agents in that group will be informed of a pending operation but only one can perform the operation. Unlike an automated operation, we cannot ensure that a manual operation will be performed as soon as the instruction for performing it is sent. Also, the data generated by a manual operation (e.g., the diagnosis result) may not be available until some time later. Through our user interface, the agent can report to the system that the operation has been done and provide its results.

5 Implementation

All components shown in Fig. 1 have been developed. We use Java, Sun Java System Application Server 9.0, Enterprise JavaBeans 3.0, the Apache jUDDI project, MySQL 5.0, and AJAX technologies to implement our system. Different from the earlier version that was demonstrated [8], the current system supports the integrated specification and processing of rules and processes/procedures using the extended specification language, user interface, and ETKnet. We present the specification language and the user interface below.

5.1 XML-Based Knowledge and Process Specification Language

In our work, we integrate the specifications of three general types of rules for capturing human and organizational knowledge and the structural constructs specified in WPDL and BPEL for defining collaborating processes and operating procedures in a single XML-based language.

5.1.1 Integrity Constraint Rule

Integrity constraints model the constraints or conditions to which data should adhere, which are dictated by requirements of an application. Since we use an object-oriented data model for modeling data associated with an event occurrence (i.e., event data), constraints can be specified on an object's attributes and relationships between attributes. Our system supports two types of integrity constraints. Constraints of the first type are called attribute constraints. They limit the acceptable values that any data attribute may have at any point in time. Such a constraint is of the form

$$x \, \theta \, n, \text{ or } x \, (\text{in / not in}) \, \{n_1, n2, \ldots n_a\}$$

where x is an object attribute, n is a value from x's domain, θ is one of the six arithmetic comparison operators ($>, >=, <, <=, =, \neq$), and $\{n_1, n_2, \ldots, n_a\}$ represents a set of enumerated values from x's domain.

Constraints of the second type are *inter-attribute* constraints. They are used to constrain relationships among attributes. The relationship can be modeled in either a mathematical way or conditional way. The former allows for so-called formula constraints, which are of the form

$$f_1(x_1, x_2, \ldots, x_b) \, \theta \, f_2(y_1, y_2, \ldots, y_c)$$

where $f_1(x_1, x_2, \ldots, x_b)$ and $f_2(y_1, y_2, \ldots, y_c)$ are mathematical formulas relating the object attributes $x_1, x_2, \ldots, x_b$, and $y_1, y_2, \ldots, y_c$, respectively. The conditional constraints are of the form

$$If \, (P_1 \, \alpha \, P_2 \alpha \ldots P_d) \, then \, (Q_1 \, \alpha \, Q_2 \, \alpha \ldots \alpha \, Q_e)$$

where $P_1 \, \alpha \, P_2 \, \alpha \ldots \alpha \, P_d$ and $Q_1 \, \alpha \, Q_2 \, \alpha \ldots \alpha \, Q_e$ are predicate expressions of the form $f_1(x_1, x_2, \ldots, x_b) \, \theta \, f_2(y_1, y_2, \ldots, y_c)$ connected by the logical operator α in {AND,OR}. Each of $P_1, P_2, \ldots, P_d$ and $Q_1, Q_2, \ldots, Q_e$ can be in its asserted or negated form. All of the attributes of entities referenced in a constraint rule are the rule's input data. The output of the rule is the truth value indicating whether the constraint is satisfied or not.

5.1.2 Derivation Rule

Derivation rules are also known as inference rules or deductive rules. They provide new data if some premises on existing data are satisfied. They are of the form

$$P \rightarrow Q, \text{ or } P => Q$$

which means that, given that the premise P evaluates to true, the conclusion Q is also true. In their general forms, P can be a set of predicates linked by logical operators AND and/or OR, and Q can be a set of predicates linked by the logical operator AND.

The attributes of entities referenced in P are the rule's input data and those referenced in Q are its output data.

5.1.3 Action-Oriented Rule

Action-oriented rules are generally expressed in the form of event–condition–action (ECA rules). The general format of this type of rule is

On E, if C then execute A

which means that, when the event E occurs, action A will be executed if condition C evaluates to true. In our work, we separate the event specification from CA specification to allow an event defined by an organization to trigger the processing of CA rule(s) defined by any other organization(s). We adopt the format of condition-action–alternative-action (i.e., If C then A else B) used in [10, 14] for action-oriented rule specification. In this rule type, attributes referenced in C as well as the input to the conditions and operations specified in A and B form the rule's input data, and the result of performing the operations in A or B is the output data.

An action is either a primitive automated/manual operation, or a structure of operations that can be decomposed and specified as a set of operation constructs. The latter is used to specify a collaborative process or operating procedure showing the execution relationships among operations. By integrating the process patterns defined in WPDL and BPEL, we include the following constructs in our specification language. These constructs can be used recursively to express more complex operation structures.

> *Sequential construct.* This construct is used to define a sequential execution order between two operations. A second operation is activated after the completion of the first operation.
>
> *Switched construct.* This construct specifies a conditional processing of operations. It is similar to the switch statement in a programming language. The conditions in branches are evaluated in the order in which they appear, and the first one whose condition is evaluated to true will be executed. If no conditional branch is taken, then a default branch is taken.
>
> *Unordered construct.* The operations listed in this construct can be executed in an unspecified order, but not concurrently. Execution and completion of all operations is required.
>
> *Selective construct.* This construct allows a number of operations to be selected randomly from a set of operations for execution in an unordered fashion. The exact number is specified in the NUM field of this construct.
>
> *Repeated construct.* This construct is used for structured looping. The loop body can be a single operation or a set of structured operations.
>
> *Split construct.* This construct is used to specify that several operations can be executed in parallel. There are two types of split: *AND-Split* and *OR-Split*. The *AND-Split* is used to specify that all successor operations can be executed in parallel after the completion of the predecessor. The *OR-Split* is used to state that a specified number of operations can start their execution in

parallel after the completion of the predecessor. These operations also have
to satisfy the data conditions specified for their processing.

Join construct. This construct is used to specify a synchronization point. There
are two types of Join constructs: *AND-Join* and *OR-Join*. In *AND-Join*, the
successor operation can only be executed after the completion of all of its predecessors. In *OR-Join*, the successor operation can only be executed after the
completion of a specified number of predecessors. The successor operation
has to satisfy the data condition of each chosen incoming transition.

5.1.4 Rule Structure

Four constructs are defined to specify the relationships between rules: *link*, *split*,
and-join, and *or-join*. A *link* relationship between two rules r and s states that rule r
must be executed before rule s, if the link is from r to s. A *split* relationship between
a rule r and a collection of rules s_1, s_2, . . ., s_n states that after executing rule r, rules
s_1, s_2, . . ., s_n can begin their execution in parallel. An and-join relationship between
a collection of rules r_1, r_2, . . ., r_n and rule s states that rule s may begin its execution
only after r_1, r_2, . . ., r_n complete theirs. An or-join relationship between a collection
of rules r_1, r_2, . . ., r_n and rule s states that rule s may begin its execution after a
specified number of rules in the collection complete theirs. We note here that the rule
structure is simpler than the operation structure because each rule represents a higher
level of specification than the primitive operations. All we need is the capability to
specify sequential, parallel, and synchronization of rules in rule processing.

5.2 User Interface

The User Interface of ETKnet provides a number of applications which are made
available to each user based on his/her role in the system. At the lowest access level,
a user is only allowed to post events, and only those events that are applicable to
his/her role in the organization. Through the interface, he/she can also report the
completion of an operation that has been assigned to him/her and provide data as
the result of the operation. The privileged system roles are those of the configuration
manager, the site administrator, the knowledge engineer, and the ontology manager.
A user may fill any subset of these roles and therein be entitled to use only those
applications associated with those roles for security reasons.

Prior to using ETKnet, a user must submit a request for access to the site administrator of his/her organization. The administrator will then review the request, make
any necessary updates to the user's profile, and select the roles that the user will play
in the system. Once the user's request has been approved, he/she may log into the
system and access the applications corresponding to his/her roles. He/She will also
be presented with a list of "alerts," which map to the manual operations that have
been assigned to him/her. Figure 4 shows three operations that have been assigned to
a user who has privileges to access all five applications. By clicking on an assigned

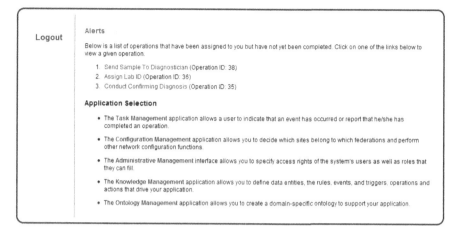

Fig. 4 Application selection and alert screen for user with all access privileges

operation, the user can commit himself/herself to performing the operation. The system will block other users, who play the same role and receive the same assignment, from viewing the same operation in their alert lists. By clicking on an application, the user can perform all the functions offered by that application.

In the task management application, a user can post a predefined event (e.g., the receipt of a sample) to indicate that the event has occurred, or report the completion of a manual operation and provide data as the result of that operation. In configuration management, a configuration administrator can access the "Federation Setup" function to specify to which federation a given organization belongs. This requires that the administrator indicates the address of the host server of the federation, as well as the database system and the e-mail server used at a collaborating site. Finally, an ontology server must be specified so that the local server's instruments can make decisions based on the domain ontology defined for the federation. In administrative management, it is the responsibility of the site administrator to approve user accounts, assign roles to users, and specify the operations that can be performed by users playing each specific role. A User Management screen displays a list of user accounts that are awaiting approval and allows the administrator to update and approve these accounts. This screen also shows all the active users in the system and allows the administrator to modify or delete them at any time. A Role Definition panel enables the administrator to add new roles or delete existing ones, and these roles can be subsequently added to the accounts of active or pending users. An Organization Registration screen makes it possible to update the contact information of a collaborating site.

In the knowledge management application, a knowledge engineer is responsible for defining data entities, events, rules, triggers, operations, operation structures, and rule structures. A Data Definition screen allows the engineer to define data entities and attributes that are relevant to a federation. These entities and attributes are mapped to the concepts of the domain ontology by using the service provided

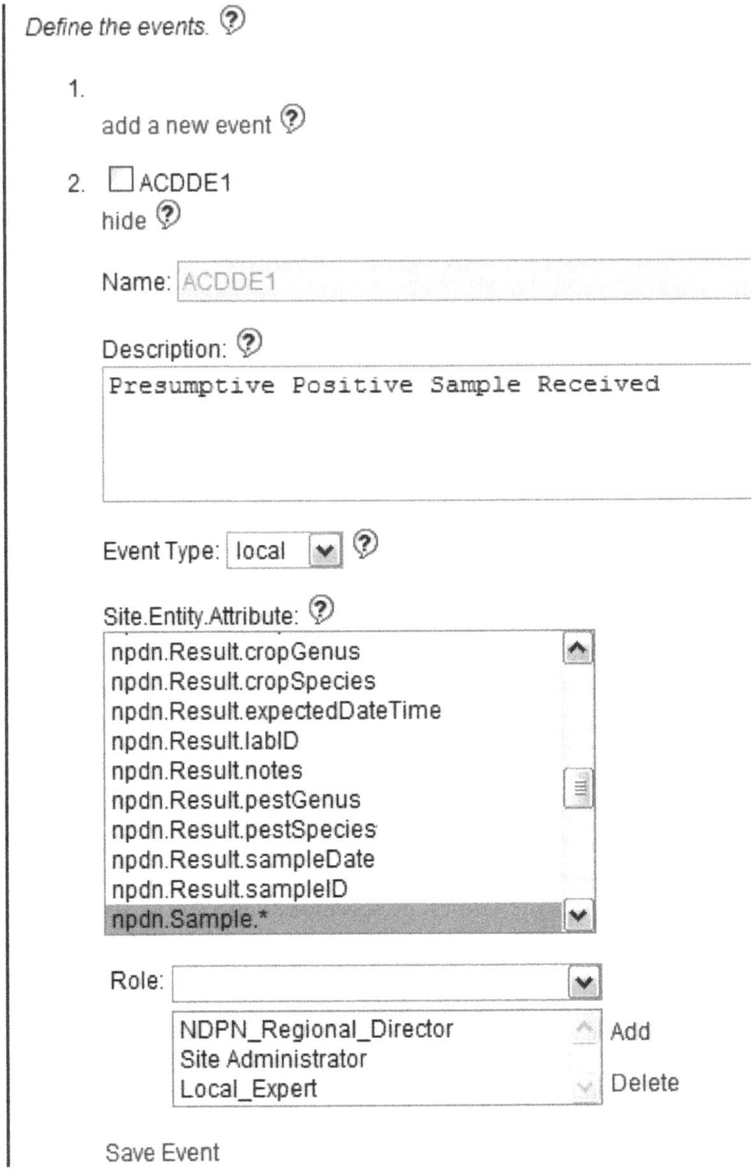

Define the events. ⑦

1.
   add a new event ⑦

2.   ☐ ACDDE1
     hide ⑦

     Name: │ACDDE1

     Description: ⑦
     ┌───┐
     │ Presumptive Positive Sample Received   │
     │   │
     │   │
     │   │
     └───┘

     Event Type: │ local ▼│ ⑦

     Site.Entity.Attribute: ⑦
     ┌───┬──┐
     │ npdn.Result.cropGenus   │▲ │
     │ npdn.Result.cropSpecies   │ │
     │ npdn.Result.expectedDateTime   │ │
     │ npdn.Result.labID   │ │
     │ npdn.Result.notes   │▤ │
     │ npdn.Result.pestGenus   │ │
     │ npdn.Result.pestSpecies   │ │
     │ npdn.Result.sampleDate   │ │
     │ npdn.Result.sampleID   │ │
     │ npdn.Sample.*   │▼ │
     └───┴──┘

     Role: │   │▼│
           ┌────────────────────────────────────┬──┐
           │ NDPN_Regional_Director   │▲ │ Add
           │ Site Administrator   │ │
           │ Local_Expert   │▽ │ Delete
           └────────────────────────────────────┴──┘

   Save Event

Fig. 5 The event definition screen allows the knowledge engineer to add new events to the system

by the Ontology Manager. An Event Definition screen shown in Fig. 5 enables the engineer to define an event by giving a name and a description that explains the meaning of the event. The engineer can then select data entities and attributes that constitute the event data. He/She can also tie the event to a role(s) so that all users who are assigned to that role(s) will be able to post the event. An event is nothing more than a name and description attached to a list of inputs; the engineer specifies these along with the roles that are allowed to post the event. The attribute list, which serves as the source for the input parameters, is selected on a separate screen where data attributes are created and attached to concepts in the ontology database.

A Rule Definition screen allows the engineer to create a rule which is one of the three types supported by the system. Depending on the type chosen, the interface will display a variety of input fields which will dynamically multiply to accommodate rules of any complexity. Figure 6 shows the screen for defining an action-oriented rule. Once a rule has been defined, it is converted into program code and wrapped as a Web service. The rule code is stored locally and the Web service is registered with the Web Service Registry at the host site.

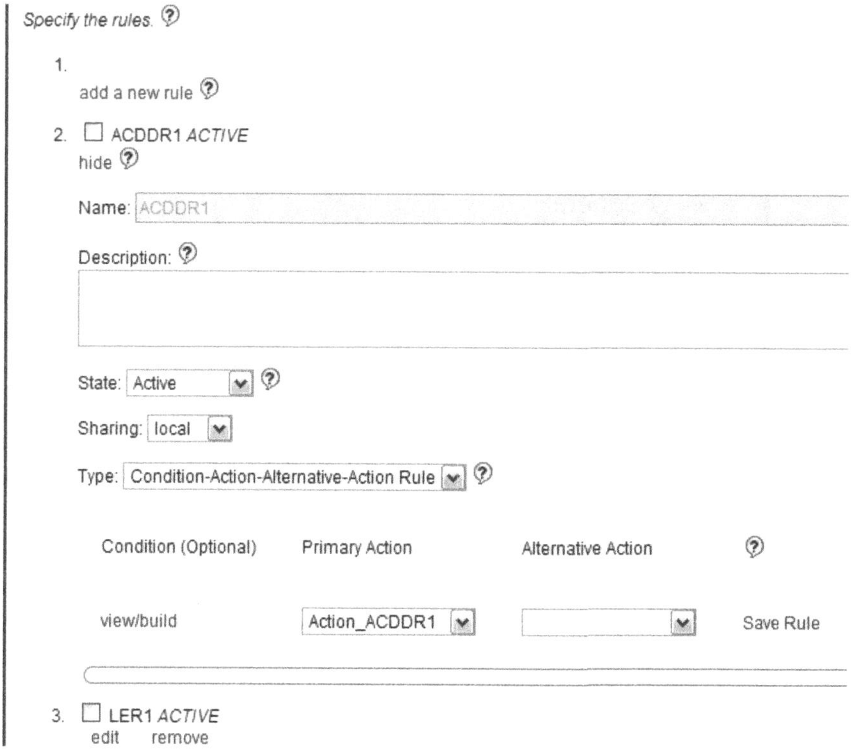

Specify the rules.

1.
    add a new rule

2. ☐ ACDDR1 *ACTIVE*
    hide

    Name: ACDDR1

    Description:

    State: Active

    Sharing: local

    Type: Condition-Action-Alternative-Action Rule

Condition (Optional)	Primary Action	Alternative Action	
view/build	Action_ACDDR1		Save Rule

3. ☐ LER1 *ACTIVE*
    edit   remove

Fig. 6 The rule definition screen allows for the creation and registration of a condition-action–alternative-action rule

A Trigger Definition screen allows for connections to be made between events and rules. Each trigger specifies which rule should be activated upon the occurrence of an event. It can also effect the execution of a rule structure defined in a Rule Structure Definition screen. An Operations screen gives the engineer a way to specify the inputs and outputs of a manual operation, as well as the role(s) of user(s) who can perform the operation. The engineer can also provide a customized message and a task checklist on this screen. The checklist is displayed to the user(s) for checking the tasks that have been performed. Another screen, operation structures, allows the

engineer to graphically compose a structure of operations that models a workflow
process or an operating procedure.

In the ontology management application, the ontology manager can view the
entities and attributes defined by the knowledge engineer and create concepts and
conceptual structures in the domain ontology for these entities and attributes. The
linkages between the names of these entities and attributes and their corresponding object ids in the ontology database are established. Figure 7 shows the display
screen.

All the display pages of the user interface are generated dynamically by a series
of Servlets and JavaServer Pages (JSPs), and are presented to the user using HTML
forms viewable through any modern Web browser.

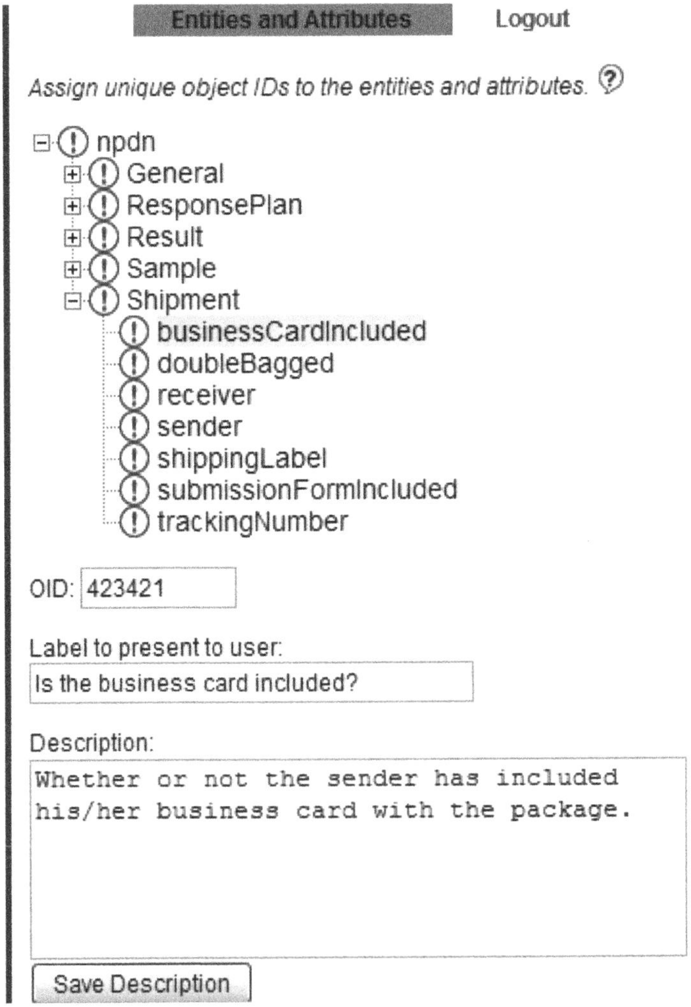

Fig. 7 The ontology manager attaches entities and attributes to concepts in the ontology database

6 Research Issues

There are several research issues that are being investigated. One involves security and trust policies. Collaborating organizations need to negotiate and establish the policies to be enforced by ETKnet. We are interested in specifying these policies in knowledge rules and rules structures so that they can be processed uniformly with other knowledge rules. There are questions concerning who can define and publish events, who can define and modify rules, who can define and modify triggers that link distributed events to distributed rules, and what should be done with those triggers wherein the event(s) specified have been deleted (i.e., considered by the collaboration federation as no longer important). Role-based access control strategy [19], the concept of membership [9], certification-based authentication [10], and PKI technologies are applicable in a federated environment. We shall adopt available technologies and, in some cases, extend them for access control and management of distributed events, rules, and triggers.

The second issue regards ontology, which was discussed in Section 3. Terms used by one organization in its specifications of events, rules, triggers, and operations, in its names for entities and attributes of event data, and in its metadata descriptions can be quite different from those used by another organization. Furthermore, different organizations may create similar but slightly different conceptual structures to represent the same concept. People searching for registered events and Web services that implement rules and rule structures will likewise face a mismatch between the terms used in their searches and the terms used to register these data and knowledge resources. In addition to the ontology defined for NPDN's SOP, we are defining a domain ontology for plant disease and pest diagnostics to be managed by the ontology management system, which will either automatically or semi-automatically deal with ontological mappings by reasoning on the underlying concepts of terms. We are also extending the Web services technology by developing an ontology-enhanced Web service registry to enable the registration, semantic discovery and invocation of application operations, rules, and rule structures.

Another issue is that of scalability. The architecture described in Section 3 allows organizations to join a federation by installing microcomputers that contain the developed software and tools at their network sites. The network is highly expandable and scalable because more computational power is added as more sharable knowledge is added to the federation. Also, since the components shown in Fig. 1 are implemented as servers, multiple federations can be accommodated: i.e., an organization can be a member of multiple federations and its servers can process different event data sets concurrently in multiple threads. However, as the number of organizations in a federation increases, the number of federations grows, and the number of event occurrences of different event types increases; the performance of the entire network can deteriorate if a centralized host is used. Event types may have to be categorized and managed by multiple hosts. Also, there are data scalability issues. For example, should an organization archive the event data sets associated with all or some of the events it subscribes? Should the distributed network archive all or some of the event data sets? How does the system determine which historical event data to keep? Can the event data associated with an event occurrence grow to a very large size because an event may trigger rules, which in turn post other events to trigger other rules? We are seeking answers to the above questions.

7 Conclusion

In this paper, we have presented our idea of capturing multi-faceted human and organizational knowledge by using three popular types of knowledge rules and rule structures. Operation structures that model collaborative processes and operating procedures are embedded in action-oriented rules, thus allowing processes/procedures and knowledge rules to be defined in a unified specification language and be processed in an integrated and interoperable fashion. The architecture of ETKnet and the user interface were described with some implementation details. We also introduced the technique of managing dynamic event data and processing distributed and heterogeneous rules to achieve knowledge sharing by converting rules and rule structures into Web services for their uniform discovery, invocation and interoperation in a Web service infrastructure. The processing technique was explained using NPDN's SOP as an example. The intended contributions of this paper are (1) introducing a multi-faceted knowledge representation for capturing organizational and inter-organizational policies, regulations, constraints, processes and operating procedures in an integrated manner, and for sharing distributed heterogeneous rules and rule structures, (2) introducing an effective mechanism for translating high-level knowledge specifications into code and wrapping them as Web services for their uniform processing and interoperation in a Web service infrastructure, thus avoiding the use of multiple rule engines and a workflow process management system, (3) presenting the architecture of an event-triggered knowledge network, and the distributed event and rule processing strategy by using agricultural homeland security as the application domain, and (4) identifying research issues for our and others' further investigation.

Acknowledgments This work is supported by NSF under Grant IIS-0534065.

References

1. Business Rules Group, "Defining business rules – what are they really?" Available: http://www.businessrulesgroup.org/first_paper/br01c0.htm
2. Bassiliades N, Vlahavas I, and Elmagarmid AK (2000) E-DEVICE: an extensible active knowledge base system with multiple rule type support. IEEE Transactions on Knowledge and Data Engineering, 12, 824–844
3. Bry F, Eckert M, Pătrânjan P, Romanenko I (2006) Realizing business processes with ECA rules: benefits, challenges, limits. In: Proceedings of International Workshop on Principles and Practice of Semantic Web Reasoning, pp. 48–62
4. Beck HW (2008) Evolution of database designs for knowledge management in agriculture and natural resources. Journal of Information Technology in Agriculture, 3(1)
5. Degwekar S, Su SYW (2006) Knowledge sharing in a collaborative business environment. In: Proceedings of the Fifth Workshop on e-Business, Milwaukee, Wisconsin. Abstract on page 60 and paper on CD

6. Degwekar S, DePree J, Beck H, Su SYW (2007) A distributed event-triggered knowledge sharing system for agricultural homeland security. In: the Proceedings of the 2007 IEEE Conference on Technologies for Homeland Security, Woburn, MA, pp. 180–185
7. Degwekar S, DePree J, Beck H, Thomas CS, Su SYW (2007) Event-triggered data and knowledge sharing among collaborating government organizations. In: Proceedings of the 8th Annual International Conference on Digital Government Research, Philadelphia, PA, pp. 102–111
8. Degwekar S, DePree J, Su SYW, Beck H (2007) A distributed event-triggered knowledge sharing system: system demo. Received the Best System Demo Award at the 8th Annual International Conference on Digital Government Research, Philadelphia, PA
9. Hayton RJ, Bacon JM, Moody K (1998) Access control in an open distributed environment. In: Proceedings of the IEEE Symposium on Security and Privacy, California, pp. 3–14
10. Johnston W, Mudumbai S, Thompson M (1998) Authorization and attribute certificates for widely distributed access control. In: Proceedings of the IEEE Seventh International Workshop on Enabling Technologies: Infrastructure for Collaborative Enterprises, pp. 340–345
11. National Plant Diagnostic Network, http://www.npdn.org
12. Miller GA, WordNet, a lexical database for the English language, Available: http://wordnet.princeton.edu/
13. OASIS, Business Process Execution Language, Available: http://www.oasis-open.org/specs/index.php
14. Object Management Group. (2001) Object Constraint Language Specification. Available: http://www.omg.org
15. Rosenberg F, Dustdar S (2005) Towards a distributed service-oriented business rules system. IEEE Third European Conference on Web Services Proceedings
16. Rouvellou I, et al. (2000) Combining different business rules technologies: a rationalization. OOPSLA 2000 Workshop on Best-practices in Business Rule Design and Implementation Proceedings
17. Rule Markup Language, Initiative, Available: http://www.ruleml.org
18. Sowa JF (2000) Knowledge Representation: Logical, Philosophical and Computational Foundations. Brooks Cole Publishing Co., Pacific Grove, CA
19. Sandhu R, Ferraiolo D, Kuhn R (2000) The NIST model for role-based access control: towards a unified standard. In: Proceedings of the First ACM Workshop on Role-Based Access Control, Berlin, Germany, pp. 47–63
20. Sirin E, Parsia B, Grau BC, Kalyanpur A, Katz Y (2004) Pellet: a practical owl-dl reasoner the 3rd International Semantic Web Conference (ISWC2004), Hiroshima, Japan
21. The National Plant Diagnostic Network Standard Operating Procedure for APHIS-PPQ Pest of Concern Scenario – General SOP, February 3, 2009
22. Ullma J (1982) Principles of Database Systems, 2nd ed. Computer Science Press, Rockville, MD
23. Ullman J (1988) Principles of Database and Knowledge-Base Systems. Computer Science Press, Rockville, MD
24. U.S. Congress (1993) Office of Technology Assessment, Harmful Non-Indigenous Species in the United States. U.S. Government Printing Office, OTA-F-565 Washington, DC
25. Widom J, Ceri S (1996) Active Database Systems, Triggers and Rules for Advanced Database Processing. Morgan Kaufmann, San Mateo, CA
26. WFMC, Workflow Management Coalition Interface 1: Process Definition Interchange Process Model, Available: http://www.wfmc.org/standards/docs/if19807m.pdf

Duplicate Work Reduction in Business Continuity and Risk Management Processes

Shi-Cho Cha, Pei-Wen Juo, Li-Ting Liu, and Wei-Ning Chen

Abstract Business continuity management (BCM) and risk management (RM) processes are very important to current organizations. The former ensures that organizations can limit losses after severe contingencies or disasters. The latter helps organizations identify potential security incidents and adopt the most cost-effective countermeasures. However, current risk management approaches or methodologies do not reflect the important differences between RM and BCM processes. Therefore, even an organization that has established RM processes may need to re-assess the risks for BCM processes. In light of this, this study proposes RiskPatrol, a risk management system that provides an integrated view of risks associated with RM and BCM processes. RiskPatrol provides an easy way for users to retain enough information for BCM while they perform risk assessment in RM processes, and vice versa. The proposed approach can improve the efficiency of establishing information security management systems by minimizing redundancies in RM and BCM processes.

Keyword Business continuity management · Disaster recovery · Emergency response · Risk management · RiskPatrol

1 Introduction

In the information era, few organizations can remain competitive without information systems. Therefore, organizations, particularly banks and other financial institutions, are requested to build their business continuity management (BCM) processes to ensure that they have the ability to limit losses in the event of severe information system contingencies or disasters. For example, Article 9 of the

Shi-Cho Cha, Pei-Wen Juo*, Li-Ting Liu**, and Wei-Ning Chen***
Department of Information Management, National Taiwan University of Science and Technology,
Taipei 106, Taiwan, ROC
e-mail: csc@cs.ntust.edu.tw; *m9609301@mail.ntust.edu.tw;
**m9609307@mail.ntust.edu.tw; ***wlchen@nsc.gov.tw

C.C. Yang et al. (eds.), *Security Informatics*, Annals of Information Systems 9,   155
DOI 10.1007/978-1-4419-1325-8_9, © Springer Science+Business Media, LLC 2010

Taiwan Regulations Governing Establishment of Internal Control Systems by Public
Companies amended in 2007 requires public companies to develop recovery plans
for failed information systems [22]. Additionally, several standards or guidelines of
information security such as ISO/IEC 17799 [12] and NIST SP800-12 [25] emphasize BCM to minimize damage to critical business processes caused by information
system failure.

However, an organization attempting to secure its information systems often
requires effective information security risk management processes[1] to optimize
the balance between information security and convenience. Essentially, risk management (RM) processes help organizations identify potential security incidents.
Organizations can then adopt appropriate countermeasures to the incidents based on
their expected loss. Moreover, an organization establishing its information security
management system (ISMS) is required by ISO/IEC 27001 [13] to implement RM
procedures.

An organization establishing a BCM process must perform risk assessment
to determine what incidents may interrupt its critical business processes [12].
Intuitively, an organization that has already established its RM process should
directly apply the results of its previous risk assessment. However, this is rarely
the case.

The example of a loss probability distribution in Fig. 1 shows that RM and BCM
processes have different focuses. Generally, BCM processes concentrate on the critical point or threshold that can interrupt critical business processes and develop
contingency plans to ensure the continuance of the critical business processes [3].
However, current risk management approaches usually focus on estimating and
mitigating the expected loss [24]. Therefore, the extreme conditions associated
with an incident exceeding the critical point may not be addressed by current risk
management approaches.

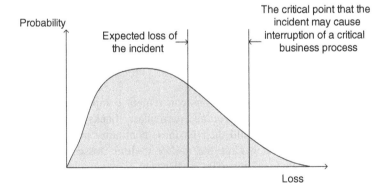

Fig. 1 Difference in focus of an incident between RM and BCM processes

---

[1] Although the concept of risk is wildly used in several domains, this chapter focuses on the risk in
information security. This chapter uses risk as information security risk.

Although over 50 risk management tools and methodologies have been developed since the early 1990s [2], few approaches integrate RM processes and BCM processes. This study therefore proposes the RiskPatrol system to support RM and BCM processes from an integrative view. Briefly, RiskPatrol provides an easy way for users to keep sufficient information for BCM while they perform risk assessment in RM processes, and vice versa. By minimizing duplication of effort, the proposed system overcomes the deficiencies of current risk management approaches.

The rest of this chapter is organized as follows: Section 2 introduces preliminary knowledge about BCM and RM. Section 3 addresses duplicate work about risk assessment for organizations to establish their BCM and RM processes. Section 4 provides an overview of the system, and Section 5 discusses its key components. Sections 6 and 7 illustrate how RiskPatrol can be applied to an organization's RM and BCM processes. Finally, Section 8 summarizes the conclusions and suggests future work.

2 Background Knowledge

2.1 Business Continuity Management

According to [19], an organization can follow the following four stages to establish its BCM processes:

- Initiation. First, an organization should develop and set its business continuity policy to provide the basis for BCM activities. The business continuity policy of an organization determines the scope of BCM within the organization, user roles and responsibilities, the review schedule of the policy, and other principles and guidelines [3, 4, 21]. The executives or senior management of an organization should approve the policy to show the importance and value of the activities. Also, sufficient money, manpower, and other necessary resources should be allocated to establish an effective BCM environment.
- Requirement and strategy. This stage includes the following activities [3, 4, 12, 19]:

  - Business impact analysis. An organization can identify its critical business processes based on its key business objectives, values, and activities. After analyzing how an interruption may impact critical business processes, the organization can determine its continuity requirement, e.g., the maximum tolerable period of the interruption of the process and the minimum requirement of the substitute process. In addition, the necessary resources for the critical business processes and associated alternative processes should also be identified.

- – Risk assessment. An organization should identify incidents that may disrupt critical business processes and should evaluate potential losses from the incidents. The BCM strategies can then be developed based on the assessment results.
- – Determining the BCM strategies. BCM strategies define how an organization should respond to the results of business impact analysis and risk assessment.

- Implementation. Depending on its BCM strategies, an organization can develop and implement contingency plans to ensure the continuance of its critical business processes. In this case, major contingency plans [26] of an organization should include the following:

  - – The *incident response* plan defines the immediate response an organization should take when incidents occur.
  - – The *disaster recovery plan* focuses on restoring critical business processes and related information systems in the original place.
  - – The *business continuity plan* defines how to establish alternative processes and information systems if critical business processes and related information systems cannot be restored within the scheduled recovery deadline.
    The plans should be tested or exercised to ensure their effectiveness. Possible approaches, including desk check, structural walk-through, simulation and complete rehearsal, are described in detail elsewhere [3, 4, 12].

- Operations management. Once previous plans are implemented, an organization must still ensure that BCM processes are standard practice. Appropriate awareness, education, or training programs are needed to embed a BCM culture within an organization. Additionally, BCM policies and plans should be updated regularly to reflect organization changes and system enhancements. Finally, self-assessment or audit is needed to identify actual and potential shortcomings of BCM processes.

2.2 Risk Management

The RM processes enable an organization to discover and assess its risks and to determine how to control or mitigate the risks as follows [7, 23, 27]:

- Risk identification. Risk identification is the process of finding out the incidents that may damage information systems and associated business processes [10]. In accordance with ISO/IEC 13335-3 [11], the following approaches can be used to identify risks. In the *baseline approach*, an organization identifies its deficiencies by comparing its current security safeguards with the minimum safeguards suggested by security standards or code of practices. For example, [15] evaluates risks of an organization according to its compliance with ISO 17799; the *informal approach* allows users to exploit their knowledge and experience when listing the risks. For example, CORAS provides a UML-based model for analyzing risks during brainstorming [16]; the *detailed risk analysis approach* involves in-depth reviews of information systems in an organization; finally, the *combined approach* is a hybrid of the above approaches. For example, users in an organization can determine the critical information systems informally. The detailed risk analysis approach is then used to identify the potential incidents affecting critical systems. For other systems, the baseline approach is used.

- Risk assessment. Risk assessment applies quantitative or qualitative approaches to predict the impacts of the identified potential incidents. Quantitative approaches usually evaluate incidents according to potential monetary loss. The most representative quantitative scheme is to calculate the annual loss expectancy (ALE) of an incident by multiplying the annual rate of occurrence (ARO) by the expected loss from the incident (or the single loss expectancy (SLE) of the incident) [24]. Instead of assigning monetary values to risks, a qualitative scheme such as OCTAVE [1], ISRAM [14], CRAMM [28], and others can be used to evaluate risks by relative levels. Generally, the qualitative scheme is easier to execute and understand by users who are not experts on security or computers than the quantitative scheme is. However, organizations can use the monetary results of quantitative scheme to calculate the return of security investment and to decide how much to insure directly [17].

- Risk treatment. After identifying and assessing a potential incident, an organization must decide how to treat the risk. Possible options include the following [12]: (1) doing nothing and accepting the risk; (2) avoiding potential incidents by changing or terminating associated actions or business processes; (3) relying on insurance or transferring the risks to other parties; and (4) applying appropriate security safeguards to mitigate the risks to an acceptable level. Several safeguards are possible [17, 20]. If more than one security safeguard option can be applied to the same potential incident, an organization can use cost–benefit analysis or other approach to optimize its security investment [8, 9, 10]. However, these issues are beyond the scope of this chapter.

- Monitoring and re-assessing the risks. The residual risks and identified acceptable risks should be regularly monitored and reviewed to ensure the accuracy and effectiveness of risk assessment and treatment, respectively [5, 13]. Additionally, an organization may need to re-assess its risks to reflect major organizational changes.

3 Duplicate Work in Business and Risk Management Processes

This section discusses duplicate work for organizations to establish their BCM and RM processes. As described before, organizations should assess their risks in both BCM and RM processes: when an organization builds its RM processes, it must find out the incidents that may damage its information systems and assess the loss expectancy of the incidents; and an organization should identify incidents that may interrupt its critical business processes when it establishes BCM processes.

Suppose that potential incidents of an organization in a specified time can be depicted in Fig. 2. We use I_{all} to represent potential incidents of an organization. If the organization assigns two teams to establish its RM and BCM processes independently at the moment, both RM and BCM team need to identify I_{all} and collect information about the incidents first. For each incident $I_i \subset I_{all}$, the RM team evaluates loss expectancy of I_i. Therefore, the RM team can find out incidents I_R that have higher loss expectancies than an acceptable value defined by the organization for further treatment. On the other hand, the BCM team recognizes incidents I_B from I_{all} by considering what incident may disrupt critical business processes of the organization.

Fig. 2 Intersection of incidents identified in RM and BCM processes

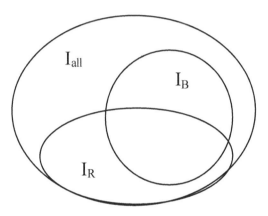

Intuitively, if the organization integrates its risk assessment activities in RM and BCM processes, it can reduce its duplicate work of incidents identification and information collection. Moreover, for each incident in $I_B \cap I_R$, the organization can adopt more effective solutions to deal with the incidents because the organization considers both expected loss and extreme conditions of the incident. At this point, this chapter proposes a systematic approach to integrate its risk assessment activities in RM and BCM processes.

4 System Overview

Generally, RiskPatrol is an information system that helps an organization to manage its risks and other related information. Figure 3 shows that RiskPatrol includes three interfaces that represent major functions. First, before evaluating the risks, users in an organization should maintain asset information through the *asset management interface*. Two major components deal with interface requests. The *asset management* component enables users to manage data for organizational assets, value and importance of the assets, incidents that may harm the assets, and so on. The

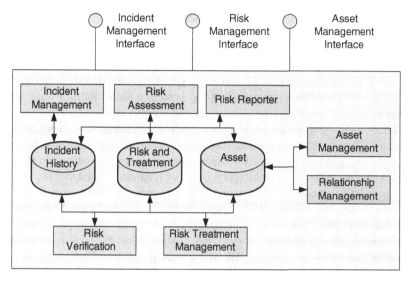

Fig. 3 System architecture

relationship management component manages relationships among assets and relationships between assets and business processes. Data for assets and relationships are stored in the *asset* database for risk identification and assessment.

The *incident management* component allows users to record data about security events in the *incident history* database through the *incident management interface*. Therefore, users can obtain information about an incident and learn from the incident.

One database and four major functional components support the *risk management interface* for risk management. RiskPatrol uses the *risk and treatment* database to manage data about risks. The *risk assessment* component provides a user-friendly means for users to identify the potential incidents and evaluate the loss expectancy of the incidents. The *risk treatment management* component helps users adopt appropriate countermeasures to deal with the risks. The *risk verification* component enables an organization to verify the accuracy and effectiveness of risk assessment and risk treatment based on past incidents. Finally, managers of an organization can access information about the risk management processes (such as the results of risk assessment and treatment) via the reports generated by the *risk reporter*.

5 Main Components

5.1 Asset and Relationship Management

The asset management component is essential for risk assessment, especially when it is not an easy task for an organization to identify and assess its potential incidents organization directly. Current risk assessment approaches usually view potential incidents to an organization as the aggregation of incidents to each asset in the organization. Asset management component reveals the incidents affecting an asset and the origin of the incidents. For example, assume an asset A_i has a set of vulnerabilities V_{A_i} and a set of threats T_{A_i} that may exploit the vulnerabilities. An incident $(T_{A_i,K}, V_{A_i,l})$ occurs when a threat $T_{A_i,K} \in T_{A_i}$ exploits a vulnerability $V_{A_i,l} \in V_{A_i}$ of the asset.

Clearly, identifying the threats and vulnerabilities of each asset in an organization is tedious. Therefore, assets are classified into different categories such as information, software, physical assets or hardware, services, users, and intangible assets in ISO/IEC 17799 [12]. For each category, the asset database is used to store data for the threats and vulnerabilities an asset of the category may have according to the common threats and vulnerabilities listed in ISO/IEC 13335-3 [11]. Consequently, potential incidents of an asset can be identified by its category. Second, the asset management component provides information about asset values to evaluate the impact of an incident in the risk assessment process as Section 5.2 demonstrates. Generally, asset values can be determined by their replacement/re-creation cost, their importance, and possible damage resulting from incidents to the assets [11, 27]. Further details are presented elsewhere.

Finally, asset ownership information is also maintained by the asset management component. Therefore, the system can control who can evaluate asset risks.

As Fig. 4 illustrates, the relationship management component maintains relationships among the major classes for asset management:

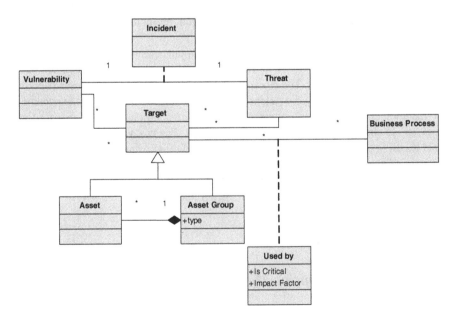

Fig. 4 Relationships among major classes for asset management

First, an organization can group similar assets into asset groups. Assets in the same asset group have similar vulnerabilities and threats for the following reasons: (1) An incident may damage several assets simultaneously. Therefore, its influence on the assets should be considered together. (2) Users can assess the risk of similar assets simultaneously for simplicity.

However, an organization can relate assets to associated business processes. For each relationship between an asset group and a business process, two attributes describe the relationship:

- The "is critical" attribute shows whether or not the asset group is critical to the business process.
- The "impact factor" attribute demonstrates the threshold that the process interrupts if the damage to the assets exceeds the threshold.

Use of the attributes is further described below.

5.2 Risk Identification and Assessment

Overall risks of an area or an organization can assumedly be viewed as the aggregation of risks to each asset or asset group in the area. For example, suppose that an area has a set of assets or asset groups $\{A_i\}$. Each asset A_i may be damaged by a set of incidents I_{A_i}. The incidents that may damage the area are equal to the unionof the incidents related to each asset or asset group I_{A_i}. While the potential incidents related to an asset can be identified based on their threats and vulnerabilities as mentioned in Section 5.1, the potential incidents to the whole area can also be determined.

RiskPatrol adopts an approach similar to that of the traditional ALE scheme [24] to evaluate the risks or potential loss of an incident based on the single loss expectancy and frequency of the incident. RiskPatrol assesses the loss expectancy of incidents as follows: Because several incidents may influence an asset, incidents related to the same asset can be assessed together rather than independently. For an incident to an asset, only the expected exposure factor must be determined. This factor represents the average proportional loss of asset value caused by the incident. Consequently, the single loss expectancy of the incident can be calculated based on the value of the asset and the expected exposure factor.

Note that the semi-quantitative assessment approach used here takes advantages of both the quantitative and qualitative risk assessment schemes. Asset value is classified into different ranges, and each range is given a user-friendly name as Table 1 illustrates. Similarly, the percentage of exposure factor with classes can be classified by range. As qualitative approaches, users can simply categorize an asset value (or its exposure factor) according to its approximate value. On the other hand, a comparative impact value can be obtained using quantitative approaches. Take Table 2 for example. An asset value can be chosen in each range of asset value (e.g., the

Table 1 Sample classification of asset values and exposure factors

Class name	Range of value	Range of percentage
D	[$0,$10K)	[0,25)
C	[$10 K,$1M)	[25,50)
B	[$1M,$100M)	[50,75)
A	[$100M,∞)	[75,100)

Table 2 An example of a mapping table about single loss expectancy

			Asset value			
			D	C	B	A
			0	10K	1M	100M
Exposure factor	D	0.25	0 (D)	2.5K (D)	0.25M (C)	25M (B)
	C	0.5	0 (D)	5K (D)	0.5M (C)	50M (B)
	B	0.75	0 (D)	7.5K (D)	0.75M (C)	75M (B)
	A	1.0	0 (D)	10K (C)	1M (B)	100M (A)

lower bound of each range), and a percentage value can be chosen in each range of exposure factor (e.g., the upper bound of each range). After multiplying these two values, the resulting value can be classified based on the asset value classes in Table 1.

The frequency classification is similar to that for asset values. For example, frequency can be categorized between two times per year and five times per year into class *B*, over five times per year into class *A*, and so forth. Further, for each class of frequency, we can select a representative frequency value. The value can be used to generate a mapping of ALE of an incident based on its single loss expectancy and frequency. The details are skipped here.

Figure 5 gives an example of how RiskPatrol evaluates the risks of an asset. The figure shows that users can see some basic information about the asset such as its custodian, type, value, name, and other descriptions. Additionally, RiskPatrol lists the potential incidents to the asset based on the type of the asset. For each incident, users can set the class of the frequency, expected exposure factor, maximum exposure factor, and time to recover from the worst case of the incident. The ALE or risk of the incident is then calculated and presented automatically.

Finally, to retain enough information about the worst case of an incident, RiskPatrol requests organizations to evaluate the maximum exposure factor that an asset may be damaged by an incident and the time needed to recover to an original state after the incident. The information is used to decide whether or not the incident interrupts a business process. Section 6 describes RiskPatrol in further detail.

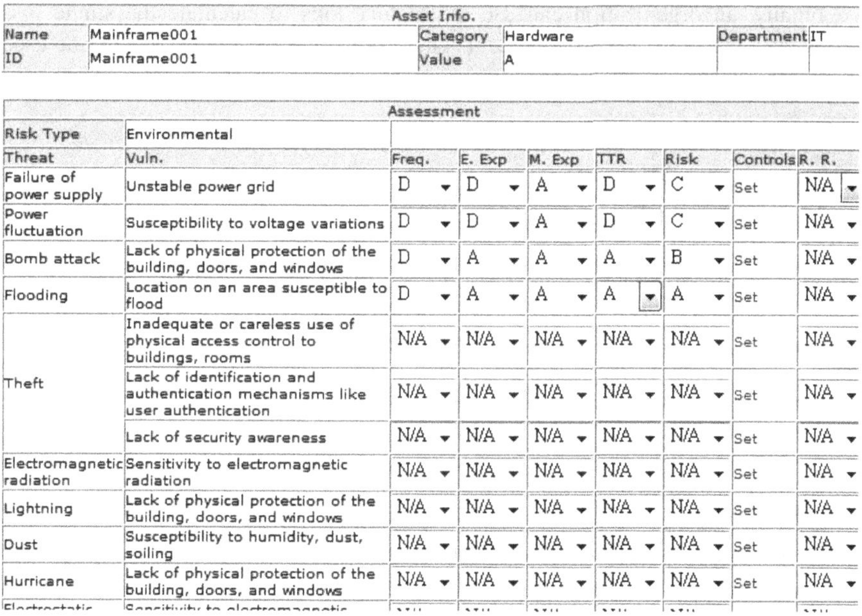

Asset Info.					
Name	Mainframe001	Category	Hardware	Department	IT
ID	Mainframe001	Value	A		

Assessment								
Risk Type	Environmental							
Threat	Vuln.	Freq.	E. Exp	M. Exp	TTR	Risk	Controls	R. R.
Failure of power supply	Unstable power grid	D ▾	D ▾	A ▾	D ▾	C ▾	Set	N/A ▾
Power fluctuation	Susceptibility to voltage variations	D ▾	D ▾	A ▾	D ▾	C ▾	Set	N/A ▾
Bomb attack	Lack of physical protection of the building, doors, and windows	D ▾	A ▾	A ▾	A ▾	B ▾	Set	N/A ▾
Flooding	Location on an area susceptible to flood	D ▾	A ▾	A ▾	A ▾	A	Set	N/A ▾
Theft	Inadequate or careless use of physical access control to buildings, rooms	N/A ▾	N/A ▾	N/A ▾	N/A ▾	N/A ▾	Set	N/A ▾
	Lack of identification and authentication mechanisms like user authentication	N/A ▾	N/A ▾	N/A ▾	N/A ▾	N/A ▾	Set	N/A ▾
	Lack of security awareness	N/A ▾	N/A ▾	N/A ▾	N/A ▾	N/A ▾	Set	N/A ▾
Electromagnetic radiation	Sensitivity to electromagnetic radiation	N/A ▾	N/A ▾	N/A ▾	N/A ▾	N/A ▾	Set	N/A ▾
Lightning	Lack of physical protection of the building, doors, and windows	N/A ▾	N/A ▾	N/A ▾	N/A ▾	N/A ▾	Set	N/A ▾
Dust	Susceptibility to humidity, dust, soiling	N/A ▾	N/A ▾	N/A ▾	N/A ▾	N/A ▾	Set	N/A ▾
Hurricane	Lack of physical protection of the building, doors, and windows	N/A ▾	N/A ▾	N/A ▾	N/A ▾	N/A ▾	Set	N/A ▾
Electrostatic	Sensitivity to electromagnetic	N/A	N/A	N/A	N/A	N/A		N/A

Fig. 5 Example of risk assessment interface

5.3 Incident Management and Risk Verification

The accuracy and effectiveness of risk assessment and risk treatment can be verified by past incidents. To achieve this purpose, the incident management component plays an important role in storing and providing appropriate information about incidents. In addition to some basic information such as an incident reference number, current status, priority, urgency level, discovery time, and closing time [18], an organization should record the impact, root cause, and related assets or asset groups related to an incident after the incident has been resolved. The information is used as follows:

Suppose an incident occurs because threat T_x exploits a vulnerability V_y of an asset or asset group A_i. We can first compare the tuple of (T_x, V_y) with potential incidents of $A_i(T_{A_i,k}, V_{A_i,l})$ where $T_{A_i,k}$ is a threat to the asset and $V_{A_i,l}$ is one of its vulnerabilities. If the tuple does not match any incident, related data about the asset can be added to the asset database.

On the other hand, while the information about assets affected by an incident is logged, the information also reveals whether or not an incident may influence several assets at the same time and whether or not the existing asset groups should be revised. Also, if some assets are often attacked by similar incidents, a new asset group can be established for the assets to reduce risk assessment costs. Conversely, the group relationship among assets can be removed when asset dissimilarity is discovered.

Finally, an organization can use the history logs to calculate the single loss expectancy and frequency by incident. Therefore, the information can later be used for asset risk assessment.

6 Integrating Risk Management and Business Continuity Management Processes

RiskPatrol supports the integration of RM and BCM processes in the following two folds: first, RiskPatrol can provide information about what incidents may interrupt a critical business process. Figure 6 demonstrates the flowchart to determine the incidents that may interrupt a process. For a process P_i with recovery time objective RTO_i, RiskPatrol first identifies the set of assets or asset groups $\{A_j\}$ that are critical to P_i based on the "is critical" attribute mentioned in Section 5.1. For each asset or asset group A_j, the threshold $TS_{i,j}$ is obtained based on its impact factor to P_i. RiskPatrol then extracts the potential incidents to the asset or asset group. For each incident I_x, the maximum exposure factor M_{I_x} and time to recovery TTR_{I_x} are

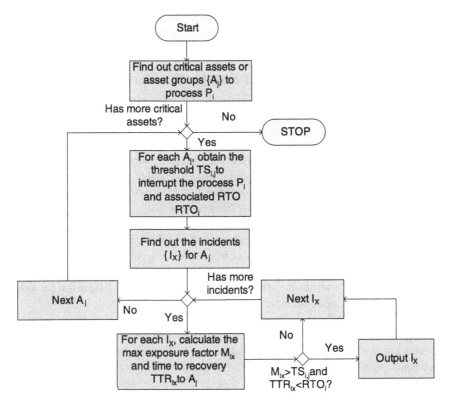

Fig. 6 Flowchart to find out incidents that may interrupt a process

compared between $TS_{i,j}$ and RTO_i, respectively. RiskPatrol selects and outputs the incidents that have higher maximum exposure factor than the threshold ($M_{I_x}>TS_{i,j}$) and smaller time to recovery than RTO of the process ($TTR_{I_x}<RTO_i$). Therefore, an organization can use the output incidents to develop its BCM strategies.

On the other hand, an organization can link its contingency plans back to the potential incidents through the risk treatment management component. Therefore, users can ensure that the organization has countermeasures for extreme cases.

7 Example

This section illustrates the RiskPatrol concept with a simple example. Assume organization X has an ERP system to support its logistic process. The ERP system is executed in a mainframe located in the basement of the organization headquarters. For simplicity, the ERP system and its associated hardware are grouped as an asset group, and their risks are assessed as a whole. After analyzing the impact of the interruption, the organization sets the recovery time objective (RTO) of its logistic process to 3 h. Restated, if the logistic process is disrupted, it should recover in 3 h.

Given the reconstruction cost and confidentiality, integrity, and availability requirement of the ERP system, the organization categorizes the system as class A based on the classes in Table 1. Table 3 lists a subset of potential incidents to the system. Because the system may still work after virus attack, the organizations evaluate the expected exposure factor of "virus" incident as B based on Table 1. The organization determines that the exposure factors of both "hard disk failure" and "flood" are A because the system fails totally after the incidents. Similarly, the frequencies of the incidents are categorized as A, C, and D based on Table 4. Risk values of the three incidents can be calculated as 375M, 0.05M, and 0.5M by multiplying the representative values in Tables 2 and 4. The classification in Table 1 shows that the three risk values of the incidents fall in class A, class C, and class C. Consequently, the organization notices the virus incident and enhances its anti-virus program to control the incident.

Table 3 Partial risks to the ERP system of organization X

	Expected exposures factor	Frequency	Risk	Maximum exposures factor	Time to recovery
Virus	B	A	A	A	D
Hard disk failure	A	C	C	A	D
Flood	A	D	C	A	A

However, the organization also assesses the maximum exposure factor and time to recovery of the three incidents. For maximum exposure factors, the organization uses the same classification scheme as expected exposure factors. For time to recovery, the organization uses the categorization in Table 5. Although all three

Table 4 Sample classification of occurrence frequency used by organization X

Class Name	Range of frequency	Representative value
D	(0,0.005]	0.005
C	(0.005,0.1]	0.1
B	(0.1,1]	1
A	$(1, \infty]$	5

Table 5 Sample classification of recovery time used by organization X

Class name	Recovery time
D	0–3 h
C	From 3 h to 1 day
B	From 1 day to 1 week
A	More than 1 week

incidents have maximum exposure factor A, only incidents of "virus" and "hard disk failure" can be recovered in RTO. However, if the headquarters of the organization is flooded, the organization needs 1 month to recover its ERP system. Consequently, the organization moves the ERP system to high floors and decides to build an overseas backup site to address the issue. As illustrated in this example, because RiskPatrol supports RM and BCM processes effectively, the duplicate work of risk assessment in RM and BCM processes can be reduced.

8 Conclusions and Future Work

The proposed risk management system provides an integrated view of risks in RM and BCM processes. Current risk management approaches or methodologies usually overlook the difference in focus between risk management processes and BCM processes. Therefore, even if an organization has established its risk management processes, it may need to re-assess the risks for BCM processes. Comparatively, RiskPatrol provides an easy way for users to retain enough information for BCM while they do risk assessment in risk management process, and vice versa. While the redundant work in RM and BCM processes can be reduced, the proposed system can hopefully correct deficiencies in current risk management approaches.

Other than a concrete implementation of RiskPatrol, some aspects require further study. First, the proposed framework focuses on information security risks. However, an organization may have several other kinds of risks, such as market risks and operational risks. Future studies may apply the proposed framework to other risk types.

Second, recall that we use semi-quantitative approach to assess the risks. In our approach, users evaluate risks by classes rather than precise value. This approach requires a trade-off between convenience and precision. More classes provide more precision but require more effort to evaluate the risks. Finding the equilibrium point may be a worthwhile topic of future studies.

Third, although similar assets can be grouped, and their risks can be assessed together, an organization may still have many assets or asset groups. Moreover, users may wish to recognize the up-to-date risks of an organization. Tools can be developed to monitor asset status and risk automatically and dynamically.

Finally, RiskPatrol currently uses only incident history to validate the effectiveness of risk assessment and treatment. The history logs can be used to predict future loss expectancy of incidents automatically or to give advice on risk assessment. For example, [6] uses the traditional linear discriminant analysis methodology to classify the potential impact and frequency of incidents. However, using past data to forecast rapid advances in information technology is questionable. Further studies are needed to explore these important questions.

References

1. Alberts CJ, Dorofee A (2002) Managing Information Security Risks: The OCTAVE Approach. Boston: Addison-Wesley Longman Publishing Co., Inc
2. Anderson AM (1991) Comparing risk analysis methodologies. In: D. T. Lindsay and W. L. Price (eds) IFIP TC11, Seventh International Conference on Information Security (IFIP/Sec'91). Elsevier, pp. 301–311
3. British Standards Institute (BSI) (2003) Guide to business continuity management. BSI Publicy Available Specification PAS56
4. British Standards Institute (BSI) (2006) Business continuity management. Code of practice. BSI Standard 25999-1:2006
5. British Standards Institute (BSI) (2006) Information security management systems – part 3: Guidelines for information security risk management. BSI Standard 7799-3:2006
6. Cha SC, Tung HW, Hsu CH, Lee HC, Tsai TM, Lin R (2005) Take Risks into Consideration while Job Dispatching, ser. IFIP International Federation for Information Processing. Springer, Boston, vol. 191/2005, pp. 1–14
7. Eloff JHP, Labuschagne L, Badenhorst KP (1993) A comparative framework for risk analysis methods. Computers & Security, vol. 12, no. 6, pp. 597–603
8. Gordon LA, Loeb MP (2002) The economics of information security investment. ACM Transactions on Information and System Security (TISSEC), vol. 5, no. 4, pp. 438–457
9. Hausken K (2006) Returns to information security investment: The effect of alternative information security breach functions on optimal investment and sensitivity to vulnerability. Information Systems Frontiers, vol. 8, no. 5, pp. 338–349
10. Hoo KJS (2000) How much is enough: a risk management approach to computer security. Ph.D. dissertation, Stanford University
11. ISO/IEC (1998) Information technology – security techniques – management of information and communications technology security – part 3: Techniques for the management of IT security. ISO/IEC TR 13335-3 Technical Report
12. ISO/IEC (2005) Information technology – security techniques – information security management systems – code of practice for information security management. ISO/IEC 17799:2005 International Standard
13. ISO/IEC (2005) Information technology – security techniques – information security management systems – requirements. ISO/IEC 27001:2005 International Standard
14. Karabacaka B, Sogukpinarb I (2005) ISRAM: information security risk analysis method. Computers & Security, vol. 24, no. 2, pp. 147–159

15. Karabacaka B, Sogukpinarb I (2006) A quantitative method for ISO 17799 gap analysis. Computers & Security, vol. 25, no. 6, pp. 413–419
16. Lund MS, Braber F, Stølen K, Vraalsen F (2004) A UML profile for the identification and analysis of security risks during structured brainstormings. SINTEF report
17. Microsoft Solutions for Security and Compliance group (MSSC), Microsoft Security Center of Excellence (SCOE) (2006) The security risk management guide v1.2. Microsoft Corporation
18. Office of Government Commerce (OGC) (2000) Service Support. TSO
19. Office of Government Commerce (OGC) (2001) Service Delivery. TSO
20. Stoneburner G, Goguen A, Feringa A (2002) Risk management guide for information technology systems. Recommendations of the NIST Special Publication 800-30
21. Swanson M, Wohl A, Pope L, Grance T, Hash J, Thomas R (2002) Contingency planning guide for information technology systems. NIST Special Publication 800-34
22. Taiwan Financial Supervisory Commission (2007) Regulations governing establishment of internal control systems by public companies
23. The Committee of Sponsoring Organizations of the Treadway Commission (COSO) (2004) Enterprise risk management – integrated framework. COSO Publications
24. U.S. Department of Commerce (1979) Guidelines for automatic data processing risk analysis. FIPS Publications 65
25. U.S. Department of Commerce (1995) An introduction to computer security: The NIST handbook. NIST Special Publication 800-12
26. Whitman ME, Mattord HJ (2006) Principles of incident response and disaster recovery. Course Technology
27. Whitman ME, Mattord HJ (2007) Management of information security, 2nd edn. Course Technology
28. Yazar Z (2002) A qualitative risk analysis and management tool – CRAMM. SANS InfoSec Reading Room White Paper